DIGITAL TAPE STANDARDS

Includes standards SEG-A, SEG-B, SEG-C, SEG-Y, and SEG-D of the SEG Committee on Digital Recording Standards and its successor, the SEG Technical Standards Committee

''Recommended standards for digital tape formats'' reprinted from GEOPHYSICS, v. 32, p. 1073-1084; v. 37, p. 36-44; v. 40, p. 344-352.

SOCIETY OF EXPLORATION GEOPHYSICISTS

ISBN 0-931830-15-X
Library of Congress Catalog Card Number: 80-52678

Society of Exploration Geophysicists
P.O. Box 3098, Tulsa, Oklahoma 74101

Published 1980, Reprinted 1981

Printed in the United States of America

CONTENTS

SEG-D—Digital field tape format standards 31

D.A. Cavers, P.E. Carroll, E.P. Meiners, C.W. Racer, L.E. Siems, M.G. Sojourner, J.L. Twombly, and J.A. Weigand

Preface

During the early years of digital recording of seismic data, numerous recording formats were put into use. Recognizing this proliferation problem, the SEG established a Committee on Digital Recording Standards in 1966. The Subcommittees on Field Tape Formats and Exchange Tape Formats developed the SEG-A, SEG-B, and SEG-X standards which were published in GEOPHYSICS in December, 1967.

Because of several new developments in field system technology (quaternary gain, instantaneous floating point amplifer, field compositing), the committee was reconstituted in 1971. The Subcommittee on Field Tape Formats developed the SEG-C standard, published in GEOPHYSICS in February, 1972.

The growing use of minicomputers in field systems prompted the committee, now renamed the Technical Standards Committee, to establish a subcommittee on Demultiplexed Field Tape Formats. This subcommittee developed the SEG-Y standard, published in GEOPHYSICS in April, 1975.

The trend toward more data channels and smaller sampling intervals, coupled with the data rate limitations of nine-track tape, led to a need for greater flexibility in format standards. The Subcommittee on Tape Formats began work on this problem in 1977. This work resulted in the SEG-D family of standard formats, published for the first time in this volume.

The establishment of digital tape format standards is a valuable function of the SEG. Now, all[1] of these important documents are collected into one publication for convenient reference.

[1]SEG-X is included because much data still exist in this format. The Technical Standards Committee, however, no longer supports SEG-X as a standard.

RECOMMENDED STANDARDS FOR DIGITAL TAPE FORMATS†

E. J. NORTHWOOD,* R. C. WEISINGER,‡ AND J. J. BRADLEY**

RECOMMENDATIONS OF THE FIELD TAPE FORMAT SUBCOMMITTEE

INTRODUCTION

Digital recording and processing of seismic information has increased very rapidly in the last five years. This advancement, although proving successful over previous techniques in many cases, has not been without problems. One of the problems faced by the contractor and major company alike has been the recording and subsequent processing of multi-formats on digital tape. The SEG, realizing this dilemma, set up a Digital Recording Standards Committee. At the second meeting of this committee in September 1966, a subcommittee was appointed to recommend a standard 9-channel field tape format. The recommendations of that subcommittee are included in this report.

The seismic record file was divided into a header block which contains constant identification information pertaining to the seismic record and the seismic data block(s) which contains the data values of the seismic channels.

Each element considered for standardization was thoroughly discussed and consensus of the subcommittee was reached before placement in the format. However, one particular feature—how should gain information be recorded—dictated the necessity of establishing two standard seismic data block formats. A common header block (Figure 1) is utilized with either type of seismic data block(s). For the purpose of discussion, the two formats are labeled Format A (Figure 2) and Format B (Figure 3).

The subcommittee would like to acknowledge the support of the Advisory Council, a group of field system manufacturers attached to the subcommittee. These organizations supplied opinions on the preliminary format designs which were most helpful in shaping these final recommendations.

DESCRIPTION OF FEATURES COMMON TO BOTH FORMATS

The following is a general description of the recommendations common to both of the proposed formats:

1. Hub size will be IBM standard.
2. The recording density will be 800 bpi with a ± 10 percent tolerance.
3. Data values are to be written in binary notation in two bytes with vertical parity odd.
4. The most significant bit of the data value will be the sign bit in the zero (bit number) bit of the first byte, and the least significant bit is in the sixth (bit number) bit of the second byte.
5. Negative numbers are to be recorded in either one's complement or in two's complement. (Format A requires data to be in one's complement.)

† These two papers are the results of work by two subcommittees of the Digital Standards Committee of the SEG. The first paper recommends standard digital tape formats for 9-track field recording of seismic data, and the second paper recommends a tape format to be used for the exchange of either field or processed data. These recommendations have a number of options to permit their acceptance by the industry. It is hoped that they will find general use by those who prefer 9-track formats. The universal use of these formats will permit a more efficient use of manpower and funds available to the industry with a resultant lowering of the cost of doing business.

These two papers should not be construed that the Committee prefers 9-track over 21-track tapes. The latter tape formats are already standardized and the Committee's efforts were directed toward the 9-track where there appeared to be a danger of proliferation of formats.

* Chairman, Committee on Digital Recording Standards (Standard Oil Company of Texas, Houston, Texas).

‡ Chairman, Subcommittee on Field Tape Formats (Gulf Oil Corporation, Houston, Texas).

** Chairman, Subcommittee on File (or Exchange) Tape Formats (Atlantic Richfield Company, Dallas, Texas).

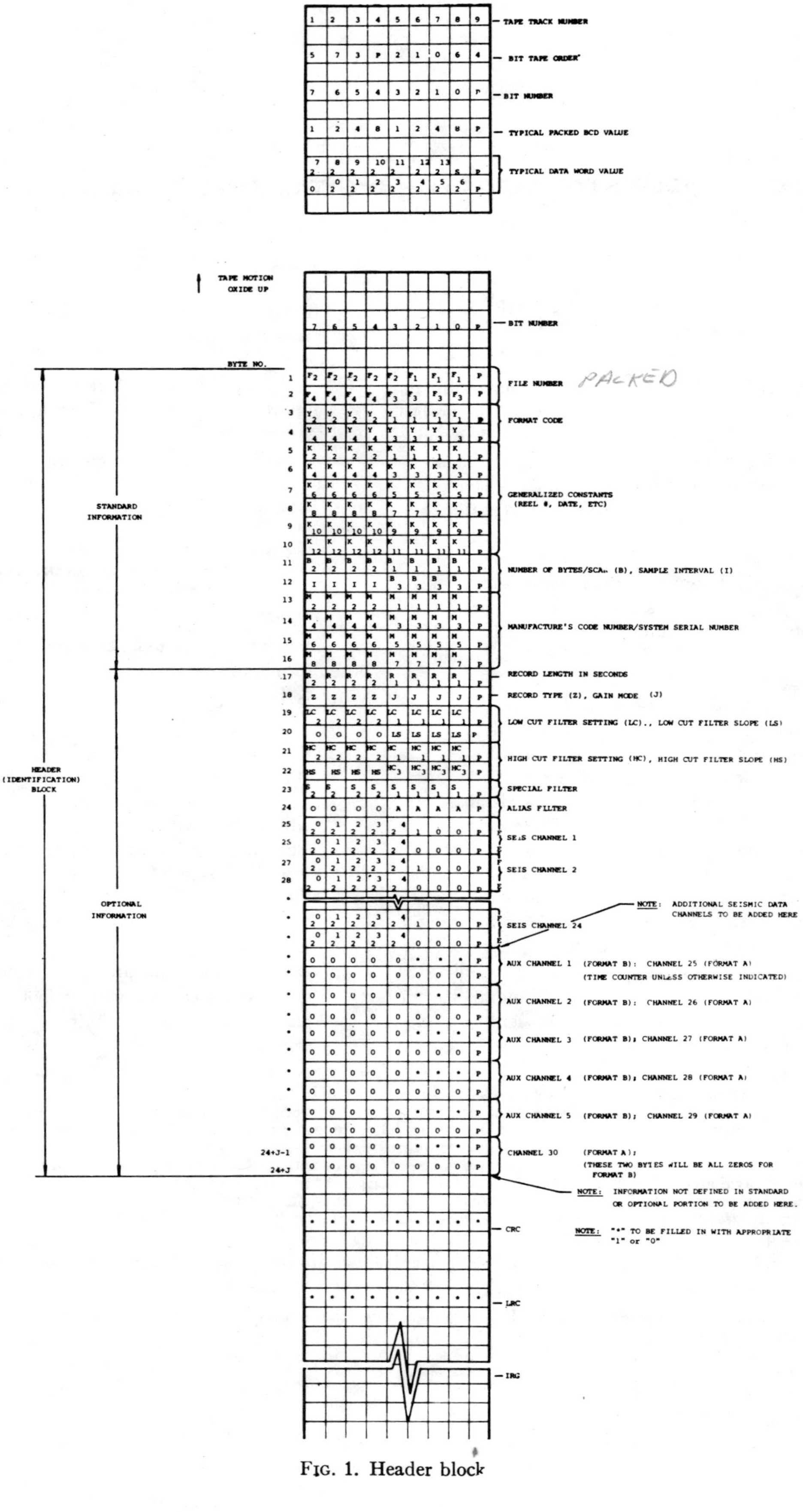

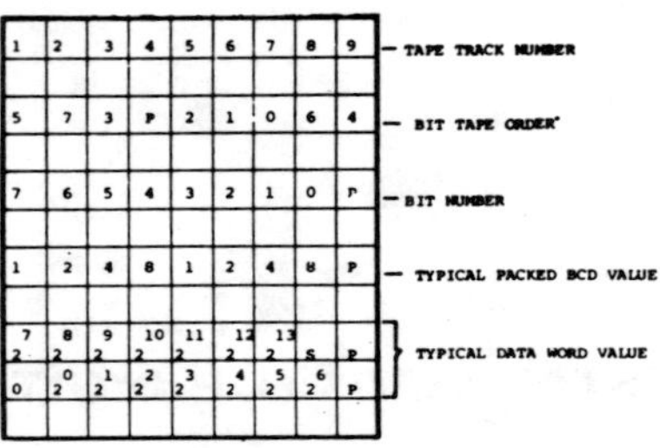

TAPE MOTION OXIDE UP

BIT NUMBER

BYTE NO.

HEADER (IDENTIFICATION) BLOCK

STANDARD INFORMATION

OPTIONAL INFORMATION

1–2 FILE NUMBER PACKED

3–4 FORMAT CODE

5–10 GENERALIZED CONSTANTS (REEL #, DATE, ETC)

11–12 NUMBER OF BYTES/SCAN (B), SAMPLE INTERVAL (I)

13–16 MANUFACTURE'S CODE NUMBER/SYSTEM SERIAL NUMBER

17 RECORD LENGTH IN SECONDS

18 RECORD TYPE (Z), GAIN MODE (J)

19–20 LOW CUT FILTER SETTING (LC)., LOW CUT FILTER SLOPE (LS)

21–22 HIGH CUT FILTER SETTING (HC), HIGH CUT FILTER SLOPE (HS)

23 SPECIAL FILTER

24 ALIAS FILTER

25–26 SEIS CHANNEL 1

27–28 SEIS CHANNEL 2

SEIS CHANNEL 24

NOTE: ADDITIONAL SEISMIC DATA CHANNELS TO BE ADDED HERE

AUX CHANNEL 1 (FORMAT B): CHANNEL 25 (FORMAT A) (TIME COUNTER UNLESS OTHERWISE INDICATED)

AUX CHANNEL 2 (FORMAT B): CHANNEL 26 (FORMAT A)

AUX CHANNEL 3 (FORMAT B); CHANNEL 27 (FORMAT A)

AUX CHANNEL 4 (FORMAT B); CHANNEL 28 (FORMAT A)

AUX CHANNEL 5 (FORMAT B); CHANNEL 29 (FORMAT A)

24+J-1, 24+J CHANNEL 30 (FORMAT A); (THESE TWO BYTES WILL BE ALL ZEROS FOR FORMAT B)

NOTE: INFORMATION NOT DEFINED IN STANDARD OR OPTIONAL PORTION TO BE ADDED HERE.

CRC

NOTE: "*" TO BE FILLED IN WITH APPROPRIATE "1" or "0"

LRC

IRG

FIG. 1. Header block

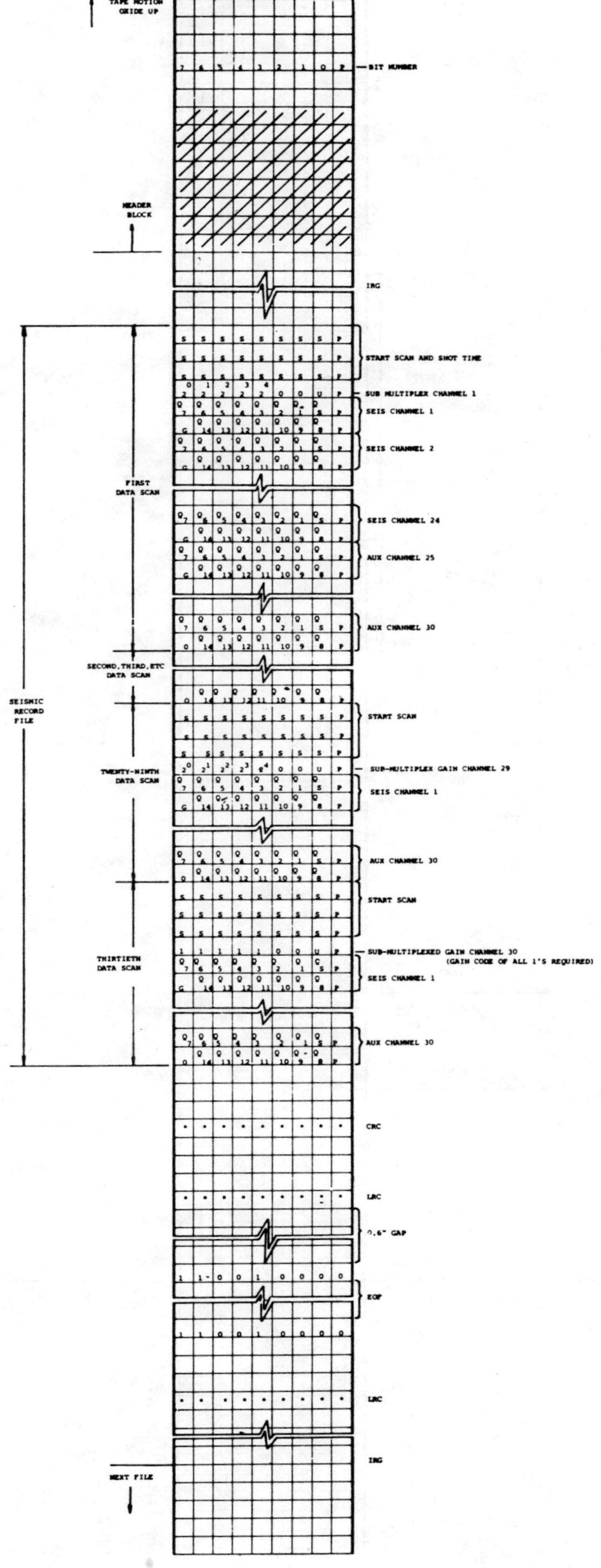

Fig. 2. Format A.

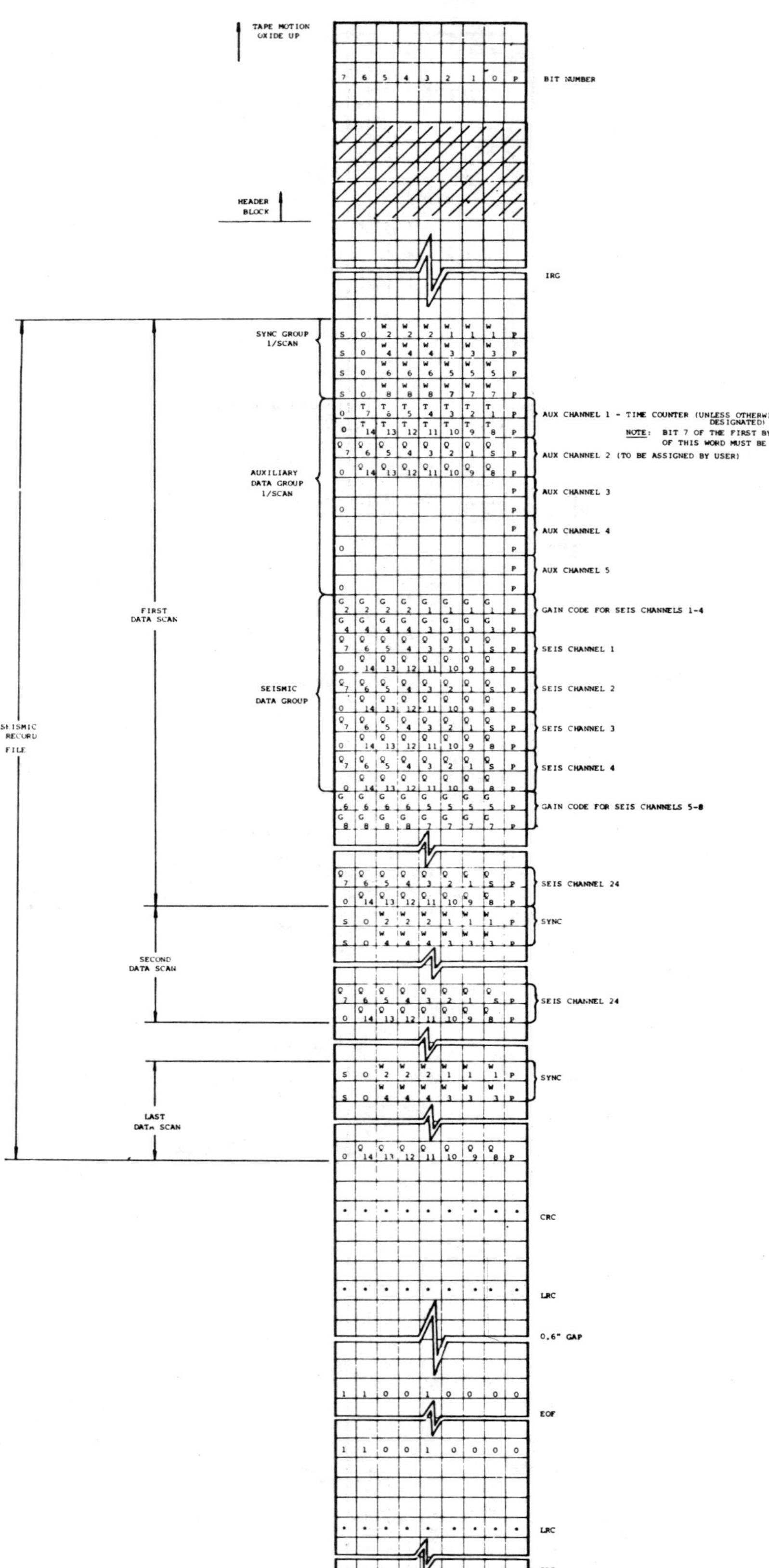

FIG. 3. Format B.

6. The NRZI method will be used to write the data words.
7. Either format may be written in a gapped or gapless format. In gapped mode of recording all seismic data blocks must contain the same number of integer data scans and be separated by standard IBM inter-record gaps.
8. Sample interval may be 1, 2, or 4 milliseconds. If longer intervals are necessary, the increments should be divisible by two, i.e., 6 or 8 milliseconds. (One millisecond not included when channel number exceeds 30 channels.)
9. A gap of at least 0.6 inch will separate the header block from the seismic data block(s).
10. An end-of-file mark will be recorded after each seismic record file with an additional end-of-file mark after the last seismic record file on the tape.
11. The formats provide twenty-four (24) seismic data channels plus auxiliary channels. In Format B the number of seismic data channels may be expanded in increments of four to any desired amount pursuant to recording capabilities.
12. Tape specifications, longitudinal parity and cyclic redundancy checks, track dimensions and numbering, and format gaps are to be compatible with IBM specifications

DESCRIPTION OF HEADER BLOCK

Each seismic record file will start with the header block shown in Figure 1. The header block is separated from its associated seismic data block(s) by an inter-record gap of at least 0.6 inch in length. The first header block on tape will begin at least 3.5 inches past the aft edge of the load-point reflective marker.

Each byte will accomodate two binary coded decimal (BCD) digits. The most significant digit (MSD) position will be in bit numbers 0, 1, 2, and 3, with respective decimal values of 8, 4, 2, and 1. The least-significant (LSD) position will be in bit numbers 4, 5, 6 and 7, with respective decimal values of 8, 4, 2, and 1.

Numbered subscripts on code letters in tape format refer to decimal digit position, e.g., Y_1—high order position of format code; Y_4—low order position for format code. The format code number 0273 will be written as follows: $Y_1=0$; $Y_2=2$; $Y_3=7$; $Y_4=3$.

Bit numbers 0, 1, and 2 of odd-numbered bytes of the pre-record gain bytes are channel identifier codes. This code is used to identify usage of each channel recorded.

Bit No.			Type Channel
0	1	2	
0	0	0	Unused channel
1	0	0	Water-break channel
0	1	0	Time-break channel
1	1	0	Gain channel
0	0	1	Seismic channel
1	0	1	Time counter
0	1	1	Uphole channel
1	1	1	Other

Bit positions 3 through 7 are reserved for pre-record gain settings. This information is in binary form with the most significant bit in bit no. 3 and the least significant bit in bit no. 7.

For each channel, the first of the two bytes is used for either the fixed gain portion of binary gain amplifiers or final gain amplifier settings.

The second byte is for the initial setting of the variable portion of binary gain amplifiers or the early gain amplifier settings.

Notes: If only the standard portion of the header block is used, the minimum length of the block must be 16 bytes even if the header information is contained in less than 16 bytes.

If a user desires to utilize any part of the optional section, then the unused bits should be written as zeros with the proper bit parity and the length of the header block will equal $J+24$ bytes.

Should information not defined in the standard or optional section be required, then these data must be recorded following the optional portion. When this extension is implemented, it will be necessary to describe these data and to obtain a new format code from the SEG Standards Committee.

DESCRIPTION OF FORMAT A—(30 CHANNELS)

Figure 2 shows the detailed arrangement of the data on the magnetic tape. Each data value of a seismic channel is recorded in two successive bytes on tape. The eight bits of a byte plus one parity check bit are recorded simultaneously on the 9 tracks of the tape at the density of 800 bpi. The data is located in the most significant 15 bits of the 2-byte "word." If a particular field recorder uses less than 15 bits for data, the most significant bit positions will be used. The 16th bit in the word is reserved for a gain change in-

Header block allocation

A. Standard Portion—16 Bytes

Byte No.	Format Letters	Description
1	F_1, F_2	4-digit file number
2	F_3, F_4	
3	Y_1, Y_2	4-digit format code assigned by SEG upon filing of format
4	Y_3, Y_4	
5	K_1, K_2	12-digit data identification constants (e.g., date, line no., reel no., etc.)
6	K_3, K_4	
7	K_5, K_6	
8	K_7, K_8	
9	K_9, K_{10}	
10	K_{11}, K_{12}	
11	B_1, B_2	Number of bytes per data scan
12 MSD	B_3	
12 LSD	I	Sample of interval in integral number of milliseconds
13	M_1, M_2	2-digit manufacturers code (See Appendix A)
14	M_3, M_4	6-digit equipment serial number
15	M_5, M_6	
16	M_7, M_8	

B. Optional Portion

Byte No.	Format Letters	Description
17	R_1, R_2	Length of record in seconds
18	J	Amplifier gain control mode 8 = binary gain 4 = programmed gain 2 = ganged AGC 1 = individual AGC 15 = other
18	Z	Type of record 8 = shot 4 = shot bridle 2 = test 1 = other
19	LC_1, LC_2	2-digit low cut filter setting
20	LS	Low cut filter slope setting in db/octave-BCD number representing slope (to calculate actual slope multiply number by 6 db/octave)
20 LSD		All zeros
21	HC_1, HC_2	3-digit high cut filter setting
22	HC_3	
22 LSD	HS	High cut filter slope setting in db/octave—BCD number representing slope (to calculate actual slope multiply number by 6 db/octave)
23	S_1, S_2	2-digit special filter setting (rejection or others)
24 MSD	A	Alias filter setting
24 LSD		All zeros
25		Pre-record gain settings and channel identifiers (fixed gain-odd numbered bytes)
26		Pre-record gain settings (early gain—even numbered bytes)
o		
o		
o		
o		
$24+J$		

Where for Format A:

J = (No. Bytes/Scan)—(No. of Bytes for Binary Gain and Sync)

For Format B:

J = (No. Bytes/Scan)—(No. of Bytes for Binary Gain and Sync) + (2 Bytes)

The two bytes were added to Format B to make the header blocks (Format A, Format B) of equal byte length.

(For a typical 24-channel system with 6 auxiliary traces, "J" will equal 60 bytes)

dication to be used with binary gain amplifiers. Auxiliary channels also use two bytes each, but the bit configuration is specialized to the types of data being recorded. This does not in any way preclude the use of auxiliary channels (except channel 30) as additional seismic channels. All the seismic channels and all the auxiliary channels are sampled and recorded once during each data-scan period.

In Format A the data portions of the seismic

record (which begins after the gap following the header block) begins with a unique start-of-scan and is coincident with the time break. In order to create a unique start-of-scan indication, it is necessary to have at least one forbidden data value. This requires four bytes, the first three of which are all 1's and the fourth byte is gain code. To avoid this same combination in other data, all 1's are ruled out. These four bytes plus the two bytes for each of the 30 data channels (seismic plus auxiliary) make up the total 64 bytes per scan used in this format.

Numerical values are written with negative numbers in 1's complement. A 1 in the sign bit indicates a negative value, while a 0 indicates a positive value. The negative zero, consisting of all 1's, is converted to the positive zero, all 0's whenever it appears. The data value consisting of all 1's is thus forbidden and as mentioned above, the start-of-scan indicator is chosen to be three successive bytes of all 1's followed by a byte (the gain code byte) in which at least one 0 is guaranteed.

In the event AVC, ganged AVC, or PGC is used, the 16th or *G* bit is not used and, if applicable, the gain trace is digitized into auxiliary channel 27. As mentioned before, with binary gain control (BGC) the 16th bit (or *G* bit) is used to indicate a factor-of-two gain change (approximately 6 db) and is recorded along with the first data sample to which the new gain applies.

In order to accomodate both increases and decreases in binary gain with the one available bit, it will be assumed that increases of gain are permitted to occur only on certain scans and that decreases of gain will be permitted only on the remaining scans. An upward change in gain is never urgent, while the necessity of making a downward change may be urgent to avoid overscaling the digitizer on large amplitudes. Any of several programs could be used, and the most popular will probably be to designate every 30th scan as an increases-of-gain scan and the remaining 29 scans would then be available for decreases-of-gain as required. In order therefore to indicate whether a given scan permits only up-changes or only down-changes to occur, a 1 is inserted in the *U* bit (see Figure 2) at the beginning of those scans in which only an up-change is permitted. A zero in the same position indicates that only down-changes are permitted for that scan. The computer program must be designed to read the *U* bit to determine whether the presence of a *G* bit represents an up or down gain change for that scan.

As a redundant check, the actual variable binary gain (or exponent value) is submultiplexed into the five least significant bits of the fourth byte of the start-of-scan indicator group. Note that the actual gain is submultiplexed by recording the gain of channel 1 in data-scan 1, the gain of channel 2 in data-scan 2, etc., and that the pattern is repeated every 30 data-scans. In order to indicate the end of the gain-submultiplex cycle, submultiplex scan, all 1's are inserted in the *Z* bits or 5 bit gain code for the 30th data scan. All 1's (31) are forbidden as a seismic data gain code.

The format requires the usual IBM cyclic-redundancy check and longitudinal redundancy check at the end of the data block. At the end of each seismic record this is followed by an end-of-file marker which comprises a 0.6-inch or greater gap and a standard end of file mark. The format requires a gap between records which is at least 0.6 inch in length and completely free of recorded bits. After the last record of interest on any tape, the end-of-file is repeated at least once to serve as an indicator to the computer that no more information follows.

SUGGESTED CHANNEL ASSIGNMENTS

Channel	Assignment
1–24	Seismic channels
25	Auxiliary or timing counter if used
26	Auxiliary or equipment quality-monitor if applicable
27	Gain control information if BGC amplifiers are not used
28	Up-hole or water-breaks
29	Time break (and water breaks if needed for split spread shooting)
30	Timing (100 Hz from an independent source)

If less than 24 seismic channels are used, channel 1 should be used with no serial order skips permitted. This, of course, does not preclude dead or omitted traces within a series of normally used channels.

FORMAT A—(MORE THAN 30 CHANNELS)

The above description is for a 30-channel version (24 seismic plus 6 auxiliary) of Format A. The general philosophy may be followed, however, to expand to a larger number of channels. The expansion should be done always keeping the seismic channels at the beginning of a multiplexer scan and the auxiliaries at the end. The submultiplex gain code and the number of bytes allowed in the header to describe the fixed and early gain should be changed appropriately. Each expansion of the format would of course require a new format code and should be filed with the SEG Standards Committee.

DESCRIPTION OF FORMAT B

Figure 3 shows the detailed arrangement of the data on magnetic tape. The Format B scan will consist of three basic groups of recorded signals that shall conform to the following minimum specifications on configuration:

Sync Group: This is a 4-byte (two-word) group consisting of a data "1" as bit 7 in track no. 2 of all bytes. All other bit positions except parity are unspecified.

Auxiliary Data Group: This is a 10-byte (five-word) group following the sync group and consisting of a data "0" as bit 7 in track no. 2 of the 1st, 2nd, 4th, 6th, 8th, and 10th byte positions to prevent a sync conflict; all other bit positions are unspecified.

Seis Data Group: This is another 10-byte (five-word) group. The first two bytes of this group define the four gain codes for data of four seismic data channels recorded in the remaining four words of the group. The gain word of the group is made of four gain codes of four bits each or a full binary range, 0 through 15. Thus, the input data shall be expressed as follows:

$$e_i = K(Q_s. Q_1 \ldots Q_{14})2^{-GGGG})$$

where e_i = input voltage,

K = scale constant (preamp gain, etc.)

$Q_s . Q_1 \ldots . Q_{14}$ is the quantized analog signal with the binary point to the left of the MSB (exclusive of sign). *GGGG* is the gain code expressed in a positive number from zero to 15.

Six of these groups will be required to scan the data for a 24-channel seismic system. If the seismic system is extended to record additional seismic gain-controlled channels, extensions in groups of four are most logical.

The following are recommended uses of unassigned bits in the three basic groups of Format B:

Sync Group:

Bits 0 through 5 may be used for recording water break counters as indicated in Figure 3.

Bits 6 in the four sync bytes may be used for different field system extensions of the number of seismic gain-controlled channels. A possible code is indicated here:

Sync byte No.	1	2	3	4	
Bit 6 Signal	0	0	0	0	indicates no extension
	1	0	0	0	indicates 1 seis data group extension
	0	1	0	0	indicates 3 seis data group extensions
	0	0	1	0	indicates 4 seis data group extensions
	0	0	0	1	indicates 6 seis data group extensions

Auxiliary Data Group:

Time Counter: (Unless otherwise designated in header block with another usage code). When a basic record time code is required, a time counter shall be recorded in the first word position of this group. This counter should be a 14-bit counter incremented at a 1 *KHz* rate. Counting will begin at the beginning of recording. The T_1 bit does not indicate sign but is considered a magnitude bit. The time break event will reset the time counter to make it count time from the time break to an accuracy of ± 0.5 ms.

Uphole/Water Depth: When uphole or water depth is required, these data will be recorded as the second word of the auxiliary data group in a fashion similar to a seismic channel (less gain code).

Time Break: When the time break is required as an analog representation, it will be recorded as the third word of the auxiliary data group similar to the uphole/water depth data. When a timing signal is required as an analog representation, it will be recorded as the fourth word of the auxiliary data group.

When any other special signal that requires no change of amplifier gain with time is to be recorded, it will be recorded as the fifth word of the auxiliary data group. Other special signals may replace previously mentioned signals in operations such as VibroSeis. For example, pilot VibroSeis signals could replace the Uphole/Time Break and/or timing word.

APPENDIX

MANUFACTURERS OF SEISMIC DIGITAL FIELD RECORDING EQUIPMENT

Code No.

01 Alpine Geophysical Associates, Inc.
65 Oak Street
Norwood, New Jersey

02 Applied Magnetics Corporation
749 Ward Drive
Santa Barbara, California 93105

03 Digital Data Systems, Inc.
7415 Hillcroft Avenue
Houston, Texas 77036

04 Dresser SIE, Inc.
Dresser Center
10201 Westheimer Road
Box 2928
Houston, Texas 77001

05 Dyna-Tronics Mfg. Corporation
5830 Star Lane
Box 22202
Houston, Texas 77027

06 Electronic Instrumentation, Inc.
601 Dooley Road
Box 34046
Dallas, Texas 75234

07 Electro-Technical Labs Division
Mandrel Industries, Inc.
A subsidiary of Ampex Corporation
6909 Southwest Freeway
Box 36306
Houston, Texas 77036

08 Fortune Electronics, Inc.
5606 Parkersburg Drive
Houston, Texas 77036

09 Geo Space Corporation
5803 Glenmont Drive
Box 36374
Houston, Texas 77036

10 Leach Corporation
405 Huntington Drive
San Marino, California

11 Metrix Instrument Co.
8200 Westglen
Box 36501
Houston, Texas 77042

12 Redcor Corporation
7800 Deering Avenue

Post Office Box 1031
Canoga Park, California 91304
13 Societe D'etudes, Recherches Et
Constructions Electroniques (SERCEL)
BP-11 44 Carquesou
Nantes, France
14 Scientific Data Systems (SDS)
1649 Seventeenth Street
Santa Monica, California 90404
15 Texas Instruments, Inc.
Box 5474
Dallas, Texas 75222

NOTE: Should additional manufacturer code numbers be required, contact the SEG Standards Committee for the assignment of these numbers.

SEG Field Tape Formats Subcommittee
J. R. Cooper
T. N. Crook
L. Gerdes
L. I. Shock
R. C. Weisinger

RECOMMENDATIONS OF THE EXCHANGE TAPE FORMAT SUBCOMMITTEE

INTRODUCTION

During the last five years the exploration segment of the petroleum industry has begun a conversion to recording and processing seismic data digitally. In the summer of 1966, the SEG set up a Digital Recording Standards Committee to study the problems for oil companies, contractors, and vendors created by a multiplicity of digital recording tape formats and to investigate the possibility of finding a format or formats acceptable to the industry as a whole. At a second meeting of this committee, a subcommittee was appointed to study and recommend a file or exchange tape format. The recommendations of that subcommittee are included in this report.

Shortly after the formation of this subcommittee, an "Exchange Tape Questionnaire" was sent to members of the geophysical industry to determine specific trends, requirements, and preferences in regard to the storage, transfer, and exchange of seismic data. On the basis of the replies to this questionnaire, this subcommittee set about the task of developing a generalized tape format suitable for filing and exchanging raw seismic field data.

In order that a file or exchange tape format be acceptable to a wide spectrum of users, it must be flexible, adaptable, comprehensive, and efficient (especially in regard to processing). The design philosophy of the proposed format includes all of these requirements plus the requirement that the format be capable of completely representing the original field data but allow for mechanical editing and demultiplexing. Provisions were made to incorporate any number of records, any number of data and information channels, all gain control systems, any length of trace, etc.; and the data and identification are variable-length records to facilitate the use of the format as an "in-house" storage and processing format if desired.

FORMAT

The format recommended for exchanging seismic data is logically divided into three sections: Header Identification, Seismic Data, and Trailer, as shown in Figure 4.

Header Identification file

The Header Identification describes the data from a line of shotpoints in a variable number of card images in narrative form. In other words, the Header does not consist of data in specific card columns in a rigid format, but the committee recommends that each card image of the Header file should contain the character C in the first card column. For example, a print-out of the first card image might be as follows:

C THIS IS LINE 537A OF THE
NORTH SEA PROSPECT

The Header section would thus consist of a series of short (80-character or byte) records separated by inter-record gaps (see Figure 4). On 7-track tape, this section would be written in BCD character code; on 9-track tape the Header would be written in EBCDIC character code. In any event, each such record would yield one line of print.

The data in this section could thus be printed and edited prior to the actual input of seismic data for processing. A listing of the Header would normally accompany a transfer tape and serve as a table of contents and summary of specifications for that line of seismic data. If two or more lines of seismic data are written on the same reel, each would have a unique Header file which would precede the data of the line described. An end-of-file tape mark follows the Header and separates it from the seismic data section.

The recommended contents of the Header Identification File are as follows:

A. Identification
Company or combine name

START OF TAPE	IDENTIFICATION HEADER FILE	EOF	SEISMIC RECORD FILE 1	EOF	SEISMIC RECORD FILE 2	EOF	//	SEISMIC RECORD FILE N	EOF	EOF	TRAILER	BLANK TAPE TO END

IDENTIFICATION HEADER FILE CI=CARD IMAGE	CARD IMAGE 1 CO. OR COMBINE	IRG	CI 2 LINE NO	IRG	CI 3 MAP OR QUAD NO	IRG	CI //	IRG	ETC	EACH CARD IMAGE WILL HAVE A C IN THE FIRST COLUMN

SEISMIC RECORD FILE	AUX. TRACE 1000	IRG	AUX. TRACE 2000	IRG	//	IRG	SEIS TRACE 1	IRG	SEIS TRACE 2	IRG	//	SEIS TRACE N

SEISMIC TRACE	TRACE IDENTIFICATION 16 TO 100 BYTES	BINARY GAIN DATA 0 TO 200 BYTES	AMPLITUDE SAMPLE //

TRACE IDENTIFICATION	RECORD SEQUENCE NUMBER	TRACE NUMBER	SP OR RECORD NUMBER	DEAD TRACE FLAG	GROUP ELEV.	GROUP DIST.	SHOT TYPE, CHARGE SIZE	ETC.

FIG. 4. Proposed standard exchange tape format.

Line number
Map number or quad name
Reel number of this set of data

B. Data type
Type of seismic source
Fold
Original data equivalent
State of processing

C. Hardware recording parameters
Equipment manufacturer, model number, and identification
Sample rate of data
Gain mode and rate of change restrictions
Recording and aliasing filters
Recording cable parameters
Specifications for the determination of time break and water-break times

D. Field recording parameters
Distances measured in feet or meters
Shothole spacing and configuration
Group spacing
Number of groups
Number of seismometers per group
Seismometer spacing and configuration
Type, frequency, hookup, and damping of seismometers

E. Data format
Number of seismic records
Number of traces per record
Number of data samples per trace
Tape format: number of tracks and packing density
Number of bytes (characters) of identification per trace (16–100)
Number of bytes (characters) of gain information per trace (0–200)
Number of bytes (characters) per amplitude sample
Negative number system (one's complement or two's complement)
Number of traces per binary gain control unit
Number of contributors per inverse gain or amplitude trace
Location and number of water-break detectors

F. Other comments
Date of start of line

Seismic data

The seismic data should be written in a demultiplexed format with the trace identification, the gain data, and the numerical values for each trace contained in a single physical magnetic tape record (see Figure 4). The traces are separated by inter-record gaps, and, if desired, groups of traces from the same seismic record may be separated from other groups by end-of-file marks. Two end-of-file marks follow the last data record for a line and precede the next Header (if more than one seismic line is written on a reel) or the Trailer for the reel. Auxiliary traces will precede seismic data traces and will receive artificial hexidecimal trace numbers. The artificial trace numbers are assigned as follows:

1,000 = gain recovery trace
2,000 = time break trace
3,000 = uphole trace
4,000 = VibroSeis pilot signal
5,000 = VibroSeis monitor
6,000 ↓ 9,000 = water-break traces

Auxiliary traces will be present only if they apply; for example, a time break trace will not be present if the time break determination is as simple as the change in value of a particular bit. The gain recovery trace will be the actual trace recorded in the field or computed from instrument settings if the data were recorded with analog equipment. This trace record is obviously dependent upon the field equipment taking the data.

Each seismic trace will have between sixteen and one hundred bytes of identification information which precedes the binary gain data and the actual seismic trace. The data to be included and a recommended sequence are as follows:

Data	*No. of Bytes Required*
Sequence number	(4)
Trace number	(4)
Shotpoint or recorded number	(4)
Dead trace flag (0 if dead, 1 if good)	(4)
Group elevation	(4)
Shotpoint-to-group distance	(4)
Coded shot type and charge size	(4)
Shot depth and elevation of shotpoint	(4)
Location of shotpoint	(8)

All additional items must be specified in the Header Identification File under "Comments." The number of bytes or characters for each item above should be specified in the Header Area.

Binary gain data will be stored compactly in about 200 bytes (characters) prior to the data values for each seismic trace. The actual number of bytes or characters used should be indicated in the Header Identification Area. The gain indicated will apply for the indicated sample number and all subsequent samples until a new gain value and sample number is entered. Thus a gain data length of 200 bytes would allow for recording the sample number and gain value for 50 gain changes.

The magnitude of the sample value will be expressed as two bytes (15 bits plus sign) in the 9-track format and in two or more characters for the 7-track format. Negative numbers should be expressed in one's or two's complement.

Trailer

The Trailer section consists of a variable number of card images with either a T (for trailer) or E (for end-of-job) in the first column. The symbol T will be used if additional seismic data for the line are on another reel. The information that should be included in this section consists of the number of complete seismic records that have been written, the number of traces from the current record that were written, and, if available, the reel number or identification tag for the tape on which the continuation of this data may be found. The symbol E will indicate the end-of-job, and then this section should summarize the data that have been written indicating the number of records and traces written, etc. The Trailer thus serves to alert the operator that a new reel must be mounted and that the present reel may be rewound.

GENERAL EXCHANGE TAPE SPECIFICATIONS

The following is a list of general recommendations for the exchange tape:

1. The magnetic tape used should be computer grade; pretested at 800 BPI and 1.5 mil Mylar base; the hub size should be IBM compatible.
2. Tape specifications, track spacing, longitudinal parity, cyclic redundancy checks, gaps, etc., are to be compatible with IBM specifications.
3. Non-Return-to-Zero (NRZ) mode should be used to write the data onto tape.
4. Packing density should be 800 bpi for nine-track tape and 556 or 800 characters per inch for seven-track tape.
5. Data values are to be written in binary mode with odd vertical parity. On nine-track tape two bytes will be used to represent each data value; on seven-track tape there should be two or three characters per data value.
6. Negative numbers should be represented in either one's complement or in two's complement.
7. The Header and Trailer data should be represented in BCD (Binary Coded Decimal) character code on seven-track tape and in EBCDIC (Extended Binary Coded Decimal Interchange Code) character code if it is on nine-track tape.

CONCLUSIONS

Though each individual company and contractor may be convinced his own format is best, unless some level of industry standardization of the exchange tape is achieved, every company will need a multitude of input and output programs to process data acquired by trading, acquisitions, mergers, and joint exploration ventures. Substantial sums of money, computer time, and programming effort can be saved if the geophysical industry will adopt a standard format for the interchange of seismic data as well as for the recording and filing of seismic data. Such standardization will save not only the initial costs of producing and checking numerous input and output programs, but also the costs of maintaining and updating these programs.

SEG Exchange Tape Format Subcommittee
James J. Bradley
George F. Cook
Sheldon E. Elliott
Tom K. Fulton
Robert B. Kerr
Carl H. Savit

GEOPHYSICS, VOL. 37, NO. 1 (FEBRUARY 1972), P. 36–44, 4 FIGS.

RECOMMENDED STANDARDS FOR DIGITAL TAPE FORMATS†

E. P. MEINERS*, L. L. LENZ‡, A. E. DALBY**, AND J. M. HORNSBY§

This paper is the result of work done by a subcommittee of the SEG Committee on Digital Recording Standards. It recommends an additional tape format for 9-track field recording of seismic data and an optional use of 1600 bpi, as well as 800 bpi, for all field formats.

INTRODUCTION

During early 1966, it became apparent that unless something was done, the number and variety of 9-track tape formats used in field recording would become unwieldy. What was needed was an industry-wide agreement to limit the number to a preferred few. To help resolve this problem, the SEG set up a Committee on Digital Recording Standards. This committee, under the chairmanship of E. J. Northwood, culminated its activities in a report (Northwood et al, 1967) which recommended two field tape formats, labeled Format A and Format B. These formats, which have been generally accepted, applied primarily to binary-gain recording. Standardization efforts with regard to 21-track formats were not considered necessary, as it was felt that the danger of proliferation was not great in this area.

During early 1971, it became apparent that owing to a number of instrument developments it was desirable again to consider the matter of standards. Among these developments were the following: (1) Instantaneous floating point amplifiers, (2) 1600 bpi recording density, and (3) possible need for larger data words to meet the requirements of field compositing.

A number of proposals were made by instrument manufacturers for new formats. These were subsequently reviewed by the present reconstituted committee. The following recommendations by the Subcommittee on Field Tape Formats is the result of considerations of these various proposals.

RECOMMENDATION

The subcommittee recommends the adoption of an IBM compatible, four bytes per sample, full-word floating point format to be designated Format C. The SEG Format A cannot accommodate the new, true floating point recording systems; and it was the sense of the subcommittee that the industry would best be served if a *single* new format could be adopted. It is further recommended that both 800 bpi and 1600 bpi be acceptable for Formats A, B, and C.

RECORDING SPECIFICATIONS

The following is a description of general recording specifications for Format C. Reference should be made to Figures 1 and 2.

1. Either the NRZI method at 800 bpi density or the phase encoded (PE) method at 1600 bpi may be used for recording. (Due to tape speed limitations of present field transports, Format C will probably be limited to

† Manuscript received by the Editor August 25, 1971.

* Chairman, Committee on Digital Recording Standards (Chevron Oil Company, Geophysical Division, Houston, Texas 77036).

‡ Chairman, Subcommittee on Field Tape Formats (Esso Production Research Company, Houston, Texas 77001).

** Member, Subcommittee on Field Tape Formats (The Superior Oil Company, Houston, Texas 77001).

§ Member, Subcommittee on Field Tape Formats (Western Geophysical Company of America, Houston, Texas 77001).

1 msec sampling of 30 channels, 2 msec sampling of 62 channels, or 4 msec sampling of 126 channels for 1600 bpi, and 2 msec sampling of 30 channels or 4 msec sampling of 62 channels for 800 bpi.)

2. The seismic record file is divided into a header block, containing identification information pertaining to the seismic record, and a data block, containing data values of the seismic channels. The first header block on a reel begins at least 3.0 inches past the trailing edge of the loadpoint reflective marker. For NRZI (800 bpi) recording, tape is erased from at least 1.7 inches before the trailing edge of the loadpoint marker to the beginning of the first header block. For PE recording, a PE identification burst is written beginning at least 1.7 inches before the trailing edge of the loadpoint marker and continuing past the trailing edge of the marker. The identification burst consists of 1600 flux reversals per inch in track P with all other tracks dc erased. The preamble of the first header block is separated from the identification burst by at least 0.5 inch of erased tape.
3. The insertion of a standard IBM interrecord gap between the header and seismic data block is optional.
4. The seismic data are to be written in gapless mode.
5. When recorded by the phase encoded method (1600 bpi), the header block, if it is separated by an interrecord gap, must be preceded by a preamble, consisting of 40 bytes of all zeros and one byte of all ones, and followed by a postamble, consisting of one byte of all ones and 40 bytes of all zeros. The seismic data block preamble may be written immediately after the minimum interrecord gap following the header block. Zero data (bit P a one and bits 0–7 all zero) are then written until time break and the first start-of-scan is written. The zero data section *must* contain an integer multiple of four bytes. The seismic data block will be followed by a postamble and end-of-file. If there is no interrecord gap between the header and the seismic data blocks, the preamble will precede the header block, and zero data (bit P a one and bits 0–7 all zero) are then written until time break and the

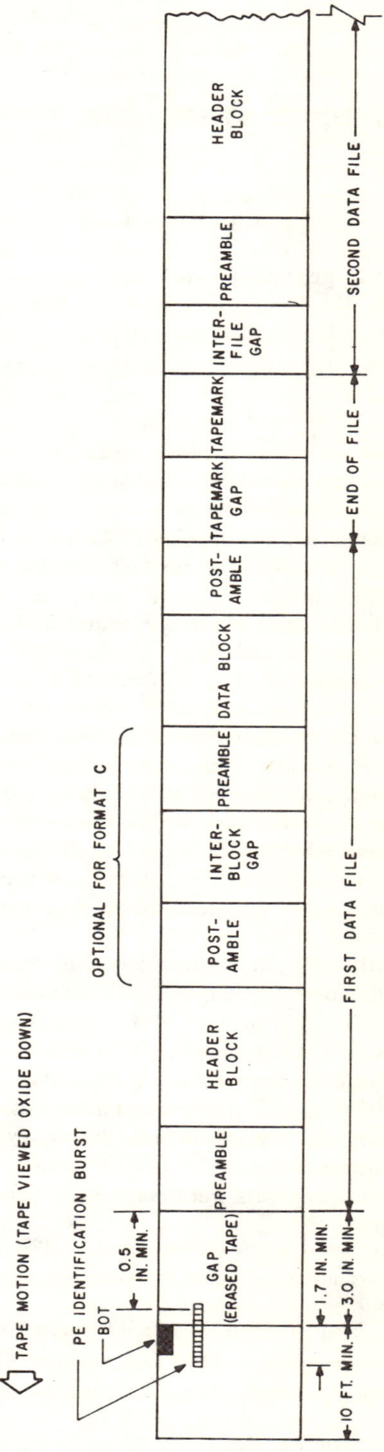

FIG. 1. General recording specifications for Format C 1600 BPI phase encoded recording. Note: Order and location of tracks on tape, direction of magnetization, and all other applicable specifications in accordance with IBM Form GA 22-6862.

first start-of-scan is written. Again the zero data section *must* contain an integer multiple of four bytes. The postamble and end-of-file will follow the seismic data block.

6. When recorded by the NRZI method (800 bpi), the header block, if it is separated from the data block by an interrecord gap, must be followed by the redundancy checks (CRC and LRC) before the interrecord gap. If the entire seismic record file is written in a gapless mode (no interrecord gap between header and seismic data), the redundancy checks and end-of-file will follow the seismic data block. Zero data (bit *P* a one and bits 0–7 all zero) are then written until time break, and the first start-of-scan is written. The zero data section *must* contain an integer multiple of four bytes. The seismic data block is followed by redundancy checks, interrecord gap, and finally an end-of-file.
7. Data values are written in eight bit bytes with vertical (byte) parity odd.
8. An end-of-file mark is recorded after each seismic record file with *at least* two end-of-file marks after the last seismic record file on the reel. The end-of-file mark consists of the 800 bpi NRZI tapemark or the 1600 bpi tapemark character, as appropriate, preceded by a gap of at least 0.5 inch of erased tape.
9. Tape specifications, location and content of redundancy check characters, track dimensions and numbering, and all other applicable specifications shall be in accordance with IBM Form GA 22-6862 entitled "IBM 2400-Series Magnetic Tape Units Original Equipment Manufacturers' Information."

DESCRIPTION OF HEADER BLOCK

The header block for SEG Format C is shown in Figure 3. It has 24 bytes of standard information identical to the first 24 bytes of the header for SEG Format A and Format B, with one exception: the LSD (bits 4–7) of byte 24 is assigned a gain constant which is common to all seismic channels. This may avoid the necessity of recording the initial gain on an individual channel basis if the initial gain is the same for all seismic channels. Data are recorded in packed BCD. Each byte accommodates two binary coded decimal (BCD) digits. The most significant digit (MSD) position is in bits 0, 1, 2, and 3, with respective decimal values of 8, 4, 2, and 1. The least signifi-

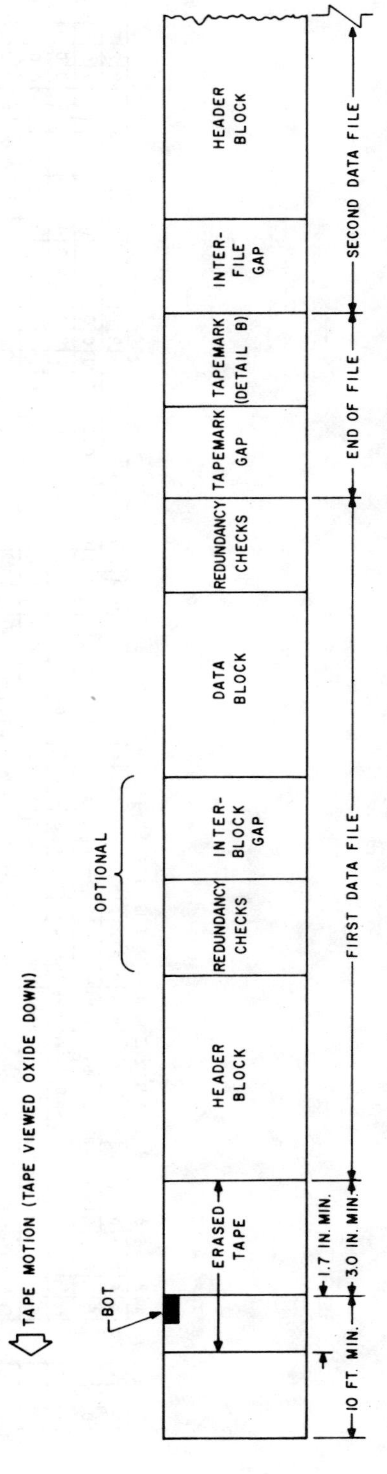

FIG. 2. General recording specifications for Format C 800 BPI NRZI recording. Note: Order and location of tracks on tape, direction of magnetization, and all other applicable specifications in accordance with IBM Form GA 22-6862.

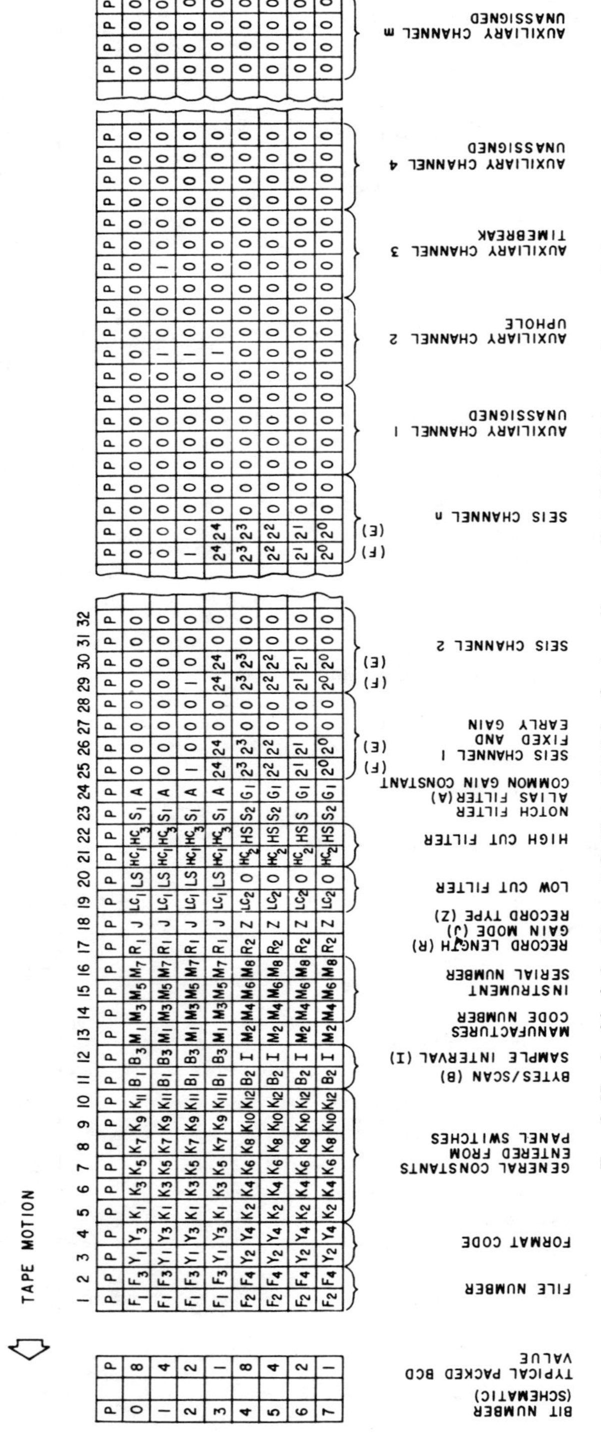

FIG. 3. Header block for Format C. Note: Additional externally supplied digital data may be recorded in the header block following byte 24 or end of the optional gain data.

cant (LSD) position is in bit numbers 4, 5, 6, and 7 with respective decimal values of 8, 4, 2, and 1.

Numbered subscripts on code letters in the format drawings refer to decimal digit position; e.g., Y_1=high order position of format code and Y_4=low order position of format code. Format code number 0273 is written as follows: $Y_1=0$; $Y_2=2$; $Y_3=7$; $Y_4=3$.

A table describing the header block entries follows:

Byte No.	*Letters*	*Description*
1	F_1, F_2	4-digit file number.
2	F_3, F_4	
3	Y_1, Y_2	4-digit format code assigned by SEG upon filing of format.
4	Y_3, Y_4	
5	K_1, K_2	12-digit data identification constants (e.g., date, line no., reel no., etc.).
6	K_3, K_4	
7	K_5, K_6	
8	K_7, K_8	
9	K_9, K_{10}	
10	K_{11}, K_{12}	
11	B_1, B_2	Number of bytes per data scan.
12 MSD	B_3	
12 LSD	I	Sample interval in integral number of milliseconds.
13	M_1, M_2	2-digit manufacturers code.
14	M_3, M_4	6-digit equipment serial number.
15	M_5, M_6	
16	M_7, M_8	
17	R_1, R_2	Length of record in seconds. Code 00 indicates continuous recording.
18 MSD	J	Amplifier gain control mode: 8=binary gain, 4=programmed gain, 2=ganged AGC, 1=individual AGC, 9=floating point gain control.
18 LSD	Z	Type of record: 8=shot, 4=shot bridle, 2=test, 1=other.
19	LC_1, LC_2	2-digit low-cut filter setting.
20 MSD	LS	Low-cut filter slope setting in db/octave, BCD number representing slope (to calculate actual slope, multiply number by 6 db/octave).
20 LSD		All zeros.
21	HC_1, HC_2	First two digits of 3-digit high-cut filter setting.
22 MSD	HC_3	Third digit of 3-digit high-cut filter setting.
22 LSD	HS	High-cut filter slope setting in db/octave, BCD number representing slope (to calculate actual slope, multiply number by 6 db/octave).
23	S_1, S_2	2-digit special filter setting (rejection or other).
24 MSD	A	Alias filter setting. 1, 2, 3, or 4 indicates sample rate of alias filter.
24 LSD	G	Common gain constant.

If it is necessary to record the initial gain of the seismic channels on an individual basis, this will begin with byte 25 and use one four byte word for each seismic channel. To maintain compatibility with the previous SEG header block, bits 3–7 of the first byte of each word are used for the fixed gain portion of either binary gain or true floating point amplifiers. This information is in binary form with the most significant bit in bit number 3 and the least significant bit in bit number 7. Bits 3–7 of the second byte are used for the initial setting of the variable gain portion of binary gain amplifiers if they are used. For true floating point amplifiers which have no initial gain settings, the second byte may be zero filled or the amplifier gain at the time of writing the header block may be recorded. For the unused bytes, zeros are recorded in bits 0–7 and one is recorded in bit *P*. Zeros are also recorded in unused bit positions of the first two bytes.

Again to maintain compatibility with the previous SEG header block, for each initial gain word, bits 0, 1, and 2 of the first byte contain the channel identifier code used to identify usage of each channel recorded as follows:

Bit no. 0	1	2	*Type channel*
0	0	0	Unused channel.
1	0	0	Water-break channel.
0	1	0	Time-break channel.
0	0	1	Seismic channel.
1	0	1	Time counter.
0	1	1	Uphole channel.
1	1	1	Other.

Should information not defined in the standard header block be required, these data may be recorded beginning in byte 25, if individual initial gain settings are not recorded, or beginning in the next byte following the initial gain settings, when initial gains are recorded. In implementing this extension, an integer multiple of four bytes must be added; and unused bits should be zero filled.

To maintain integrity of the start-of-scan code, no data may be recorded in the header in a form which might produce three successive bytes of all ones.

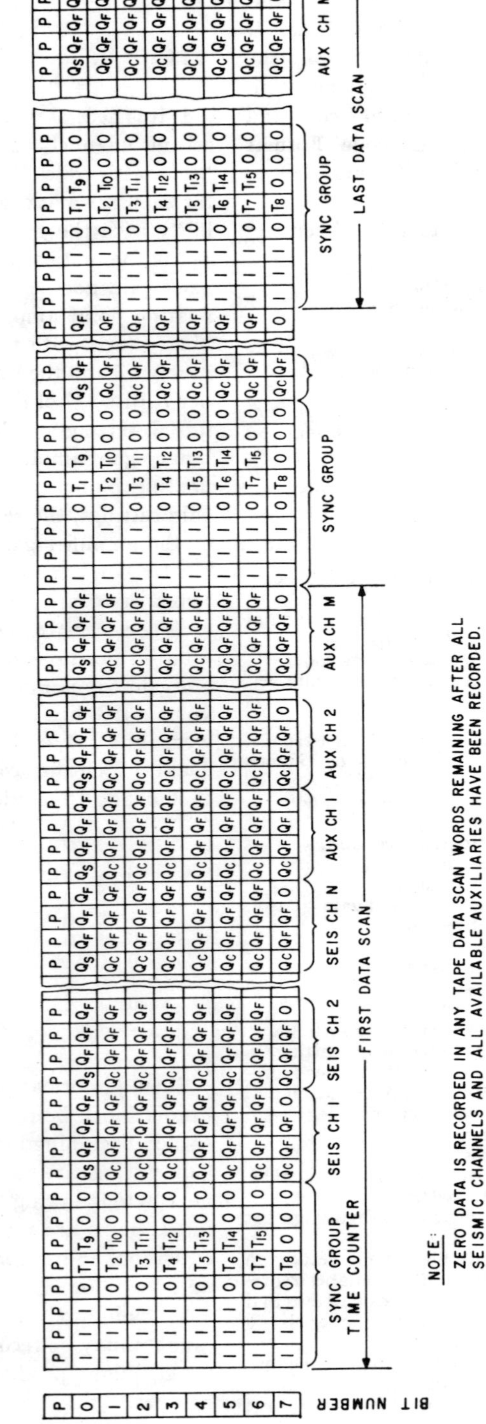

Fig. 4. Data block for Format C. Note: Zero data are recorded in any tape data scan words remaining after all seismic channels and all available auxiliaries have been recorded.

DESCRIPTION OF DATA BLOCK

A detailed drawing of the data block is shown in Figure 4. It begins with an eight byte synchronization group which is recorded preceding seismic channel 1 at the start of each data scan. The first synchronization group recorded is coincident with the time-break.

To maintain compatibility with SEG Format A, the start-of-scan indication is provided by three successive bytes of all ones recorded in bits 0–7 of the first three bytes, and all zeros recorded in bits 0–7 of the fourth byte of the sync group. A recommended 15 bit binary time counter (counting time in milliseconds from the time-break) may be recorded in the fifth and sixth bytes of the sync group (T_1–T_{15}). The seventh and eighth bytes of the sync group contain all zeros in bits 0–7. There will be a single start-of-scan indication per scan regardless of the number of channels being recorded.

Each data value of a seismic channel is recorded in four successive bytes, using IBM compatible floating point notation as defined in IBM Form GA 22-6821, entitled "IBM System/360 Principles of Operation." The four bytes form a 32 bit word consisting of a sign bits Q_S, a seven bit exponent, or characteristic, Q_C, and a 24 bit fraction Q_F.

Q_S	Q_C	Q_F
0	1 7	8 31

Q_S indicates signal polarity and is a one for negative. Q_C signifies a power of 16 expressed in excess 64 binary notation, allowing both negative and positive powers of 16 to be represented by a true number. Q_F is a six hexadecimal digit (24 binary bit) number with a radix point to the left of the most significant digit. The significance of the fraction is

$$2^{-1} + 2^{-2} + 2^{-3} \cdots + 2^{-23} + 2^{-24}.$$

The data value represented by a floating point number is

$$Q_S \times 16^{(Q_C - 64)} \times Q_F.$$

Consider a typical field recorder which uses a 15 bit analog-to-digital converter to measure channel output. The converter output is a sign bit and 14 binary magnitude bits, with the least significant magnitude bit weighted 2^{-14} (or 0.00006103515625) times the digitizer reference voltage, and the most significant bit weighted 2^{-1} (or 0.5) times the reference voltage. When the converter measures a positive full scale signal, $(1.0-2^{-14})$ times the reference voltage, its output is

$$\underset{\text{SIGN}}{\underline{0}} \quad \underset{\text{MAGNITUDE}}{\underline{1111\ 1111\ 1111\ 11}}.$$

Thus, if the recorded data value is considered as a multiplier of the digitizer reference voltage, the full scale value recorded would be

$$\underset{Q_S}{\underline{0}}\ \underset{Q_C}{\underline{1000000}}\ \underset{Q_F}{\underline{1111\ 1111\ 1111\ 1100\ 0000\ 0000}}.$$

For the special case where the reference voltage can be expressed as a power of 16, the reference voltage can be incorporated in Q_C. Multiplying the data value by 4096, or 16^3, yields the signal amplitude in millivolts if the reference voltage is 4096 mv. Thus, the recorded representation in millivolts in Format C would be

$$\underset{Q_S}{\underline{0}}\ \underset{Q_C}{\underline{1000011}}\ \underset{Q_F}{\underline{1111\ 1111\ 1111\ 1100\ 0000\ 0000}}.$$

Note that the fraction is a true number regardless of signal polarity. A negative full-scale signal would be recorded in one's complement code in SEG Formats A and B as

$$\underset{\text{SIGN}}{\underline{1}} \quad \underset{\text{MAGNITUDE}}{\underline{0000\ 0000\ 0000\ 00}}.$$

The corresponding representation in Format C is

$$\underset{Q_S}{\underline{1}}\ \underset{Q_C}{\underline{1000000}}\ \underset{Q_F}{\underline{1111\ 1111\ 1111\ 1100\ 0000\ 0000}}.$$

In this and all following examples, the value recorded in millivolts, for a reference voltage of 4096 mv, can be obtained by adding 3 (binary 0000011) to Q_C. Also note that the fraction is normalized; that is, the most significant digit (four most significant bits) *cannot* be zero unless the entire fraction is zero.

If the converter input is equal to the least significant bit, the converter output is

$$\underset{\text{SIGN}}{\underline{0}} \quad \underset{\text{MAGNITUDE}}{\underline{0000\ 0000\ 0000\ 01}}.$$

The recorded representation is:

$$\underset{Q_S}{\underline{0}}\ \underset{Q_C}{\underline{0111101}}\ \underset{Q_F}{\underline{0100\ 0000\ 0000\ 0000\ 0000\ 0000}},$$

which is obtained by left-shifting the magnitude

12 places (multiplying by 16^3) and subtracting 3 from the characteristic (dividing by 16^3).

For simplicity, the previous examples have dealt with channel output level. In a variable gain recording system, of course, the channel input, rather than the output, is the data of interest. In Format C, the recorded data value is the quotient of the channel output divided by the channel gain, or the input signal referenced to the digitizer reference voltage. The Format C representation of signals producing positive full scale outputs at various channel gains is shown below:

6 db 0 1000000 0111 1111 1111 1110 0000 0000

18 db 0 1000000 0001 1111 1111 1111 1000 0000

24 db 0 0111111 1111 1111 1111 1100 0000 0000

72 db 0 0111101 1111 1111 1111 1100 0000 0000

96 db 0 0111100 1111 1111 1111 1100 0000 0000.

Either the variable portion of the channel gain alone or the total channel gain (fixed plus variable) may be used as the divisor. The total gain should always be used when all channels do not operate at the same fixed gain. When the total gain is used, recording of the fixed and early gains in the header block is optional for Format C.

For the special case where the channel output is zero, the recorded representation is always

0 0000000 0000 0000 0000 0000 0000 0000

regardless of channel gain.

Auxiliary channels are also recorded in four bytes each, with the bit configuration specialized to the types of data being recorded. Where the auxiliary channel data are a digitized analog signal, data are recorded in the same manner as seismic channels. If the auxiliary channel contains digital data, the digital data are left justified, so that the most significant bit is in bit 0 of the first byte of the word.

Although the fractional portion of the floating point notation can accommodate 24 bits significance, the examples above show that with no field summing, and with a 15 bit converter, there will be at least seven significant bits unused. Unused bits should be filled with zeros. To maintain integrity of the start-of-scan, bit 7 of the fourth byte of each seismic or auxiliary channel word must always be zero; and in the floating point format the largest negative number that can safely be written is $-5.9 \times 10^{+75}$. If an auxiliary channel is recorded using some notation other than the floating point notation, care must be taken that it cannot produce three successive bytes of all ones.

The assignment of channels as seismic or auxiliary channels is completely flexible, provided the seismic channels are sampled sequentially, beginning with channel 1 and followed by the auxiliary channels.

Suggested channel assignments for the 30 channel system are as follows:

Channel	*Assignment*
1–24	Seismic channels.
25	Auxiliary.
26	Auxiliary or equipment monitor.
27	Auxiliary.
28	Uphole or water breaks.
29	Time break.
30	Timing (100 hz from an independent source).

EXPANSION OF FORMAT C TO MORE THAN 30 CHANNELS

For a 30 channel system, Format C utilizes 128 bytes per data scan. Expansion of the format beyond 30 channels will be in integer multiples of 128 bytes per data scan.

RECOMMENDED STANDARDS FOR 1600 BPI RECORDING OF SEG FORMATS A & B

The SEG Subcommittee on Field Tape Formats regards 1600 bpi phase encoded recording as essentially independent of any particular data format. For this reason, the committee recommends that both SEG Formats A and B allow either 800 bpi NRZI or 1600 bpi phase encoded recording, and that these formats remain unchanged in all other respects. Figures 1 and 2 illustrate general specifications of both recording modes.

The following is a description of 1600 bpi recording tape format:

1. A PE identification burst is written beginning at least 1.7 inches before the trailing edge of the loadpoint reflective marker and continuing past the trailing edge of the marker. The identification burst consists of 1600 flux reversals per inch in track *P* with all other tracks dc erased.
2. A seismic record file is divided into a header block, containing constant identification information pertaining to the seismic record, and a seismic data block(s), containing data values of the seismic channels. A gap of at least 0.6 inch of erased tape separates the header block from the seismic data block(s).

The first header block on tape begins at least 3.0 inches past the trailing edge of the loadpoint reflective marker.

3. Each block, header or data, must be preceded by a preamble consisting of 40 bytes of zero-bits in all tracks and one byte of all one-bits, and followed by a postamble consisting of one byte of all one-bits and 40 bytes of zero-bits in all tracks. The preamble of the first header block is separated from the identification burst by at least 0.5 inch of erased tape. The data block preamble must be written immediately after the minimum interrecord gap following the header block. Zero data (bit *P* a one and bits 0–7 all zero) may be written until the first start-of-scan is written.
4. An end-of-file mark is written after each seismic record file, with at least two end of file marks after the last seismic record file on the tape. The end-of-file mark is preceded by at least 0.5 inch of erased tape.
5. Tape specifications will be in accordance with Form GA 22-6862.

ACKNOWLEDGMENTS

The various proposals were originally presented at a meeting attended by representatives from a number of oil companies, contractors, and instrument manufacturers. A preliminary version of the present recommendation was sent to each of these for comment. We gratefully acknowledge the support and helpful suggestions received.

REFERENCE

Northwood, E. J., Weisinger, R. C., and Bradley, J. J., 1967, Recommended standards for digital tape formats: Geophysics, v. 32, p. 1073–1084.

RECOMMENDED STANDARDS FOR DIGITAL TAPE FORMATS[†]

K. M. BARRY,[*] D. A. CAVERS,[‡] AND C. W. KNEALE[§]

INTRODUCTION

Recently a new demand for demultiplexed formats has arisen in the seismic industry due to the utilization of minicomputers in digital field recording systems and because of a growing need to standardize an acceptable data exchange format.

In 1973 a subcommittee of the SEG Committee on Technical Standards was organized to gather information and develop a nine-track, ½-inch tape, demultiplexed format for industry acceptance. Guidelines set for this new format were based on prior work and on the SEG Exchange Tape Format (Northwood et al, 1967). As a result of the subcommittee's effort based on suggestions from industry personnel, the following demultiplexed format recommendations are made.

The present SEG Exchange Tape Format is often referred to as the SEG "Ex" Format. Because of this, it is recommended that the new demultiplexed format be designated the "SEG Y Format." The Technical Standards Committee has elected to withdraw support of the SEG "Ex" Format."

The SEG Y Format was developed for application to computer field equipment and in the present data processing center with flexibility for expansion as new ideas are introduced. Current information for standardization is placed in the "fixed" portion of the format, while new ideas can be added to the unassigned portions later as expansion becomes necessary.

It is assumed that this format will accommodate the majority of field and office procedures and the techniques presently utilized.

FORMAT SPECIFICATION

The following general information describes the recommended demultiplexed format (Figure 1):

(1) Tape specifications, track dimensions and numbering, and all other applicable specifications shall be in accordance with IBM Form GA 22-6862 entitled "IBM 2400-Series Magnetic Tape Units Original Equipment Manufacturers' Information".

At the present time, IBM has proposed an American National Standard for the 6250 CPI group coded recording format. Should this format be used within the geophysical industry, the applicable IBM specifications would apply. The additicnal formating required by this proposed method is a function of the hardware and thus becomes transparent to the user.

(2) Either the NRZI encoded data at 800 bpi density or the phase encoded (PE) data at 1600 bpi density may be used for recording.

(3) All data values are written in two's com-

† This report is the work of the Subcommittee on Demultiplexed Field Tape Formats of the SEG Technical Standards Committee. Manuscript received by the Editor October 7, 1974.

* Subcommittee Chairman, Teledyne Exploration, Houston, Tex. 77036.
‡ Subcommittee Member, Gulf Research & Development Co., Houston, Tex. 77036.
§ Subcommittee Member, Texas Instruments Inc., Houston, Tex. 77001.

Nine Track, 800 bpi NRZI or 1600 bpi Phase Encoded (PE) Demultiplex (Trace Sequential) Format

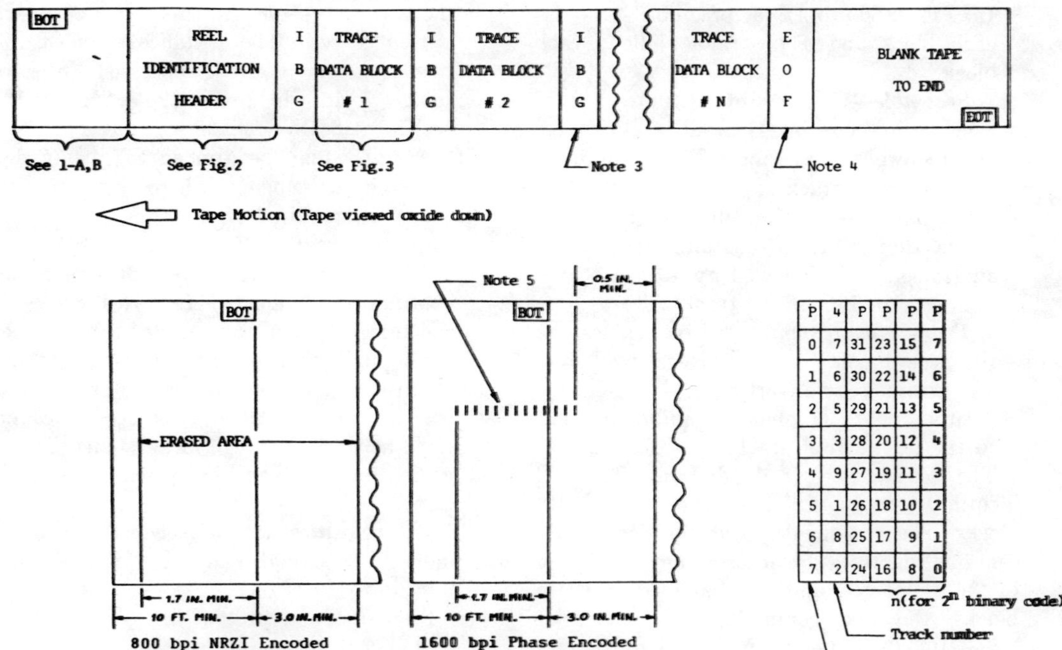

1-A 1-B 1-C

NOTES:

1. Preamble-Preceeds each of the 45 blocks within the reel identification header and each trace data block when 1600 bpi PE is used. Consists of 40 all-zero bytes followed by one all-ones byte.

2. Postamble-Follows each of the 45 blocks within the reel identification header and each trace data block when 1600 bpi PE is used. Consists of one all-ones byte followed by 40 all-zero bytes.

3. Interblock Gap (IBG)-Consists of 0.6" nominal, 0.5" minimum.

4. End of file (EOF)-Consists of an IBG followed by:
   (a) PE tape mark having 80 flux reversals at 3200 fci in bit numbers P,0,2,5,6, and 7. Bits 1,3, and 4 are dc-erased.
   or (b) NRZI tape mark having two bytes with one bits in bit numbers 3,6, and 7 separated by seven all-zero bytes.

5. PE Identification Burst-Consists of 1600 flux reversals per inch in bit number P; all other tracks are erased.

FIG. 1. Recommended demultiplexed format.

plement except the 32-bit floating point format, Figure 3-A, which is sign, characteristic, and fractional part.

(4) Data values are written in eight-bit bytes with vertical parity odd.

(5) Definitions:

(a) *Interblock gap (IBG)*—Consists of erased tape for a distance of 0.6 inches nominal, 0.5 inches minimum.

(b) *End of file (EOF)*—Consists of the 800 bpi NRZI tape mark or the 1600 bpi tape mark character, as appropriate, preceded by a standard IBG.

(c) *Erased tape*—The tape is magnetized, full width, in a direction such that the rim end of the tape is a north seeking pole. The read-back signal from such an area shall be less than 4 percent of the average signal level at 3200 flux reversals per inch.

(d) *PE identification burst*—Consists of 1600 flux reversals per inch in bit number *P* with all other tracks DC erased. This burst is written beginning at least 1.7 inches before the trailing edge of the beginning of tape (BOT) reflective marker and continuing past the trailing edge of the marker, but ending at least 0.5 inch before the first block.

(e) *Block*—Continuous recorded information,

preceded and followed by a standard IBG. In PE (1600 bpi), a preamble precedes each block and a postamble follows each block.

(f) *Preamble*—Consists of 41 bytes, 40 of which contain zero bits in all tracks; these are followed by a single byte containing one bits in all tracks.

(g) *Postamble*—Consists of 41 bytes of which the first byte contains one bits in all tracks; it is followed by 40 bytes containing zero bits in all tracks.

(h) *Two's complement*—Positive values are the true binary number. Negative values are obtained by inverting each bit of the positive binary number and adding one (1) to the least significant bit position.

(6) The seismic reel is divided into the reel identification header and the trace data blocks. The reel identification header section contains identification information pertaining to the entire reel and is subdivided into two blocks, the first containing 3200 bytes of EBCDIC card image information (equivalent of 40 cards) and the second consisting of 400 bytes of binary information. These two blocks of the reel identification header are separated from each other by an IBG. Each trace data block contains a trace identification header and the data values of the seismic channel or auxiliary channels. The reel identification header and the first trace data block are separated by an IBG.

(7) Each seismic-trace data block is ungapped and is written in demultiplexed format with each trace data block being separated from the next by an IBG. The last trace data block on the reel is followed by one (or more) EOF.

(8) When recorded 800 bpi (NRZI), the first block of the reel identification header begins at least 3.0 inches past the trailing edge of the BOT marker.

(9) The following conventions pertain to the reel and trace identification headers:

(a) All binary entries are right justified. All EBCDIC entries are left justified.

(b) All times are in milliseconds with the exception of the sample interval which is designated in microseconds.

(c) All frequencies are in hertz.

(d) All frequency slopes are in db/octave.

(e) All distances (lengths) are in feet or meters, and these systems are not mixed within a reel. The distance or measurement system used is specified in card image 7 and in bytes 3255-3256 of the reel identification header.

(f) A scaler may be applied to certain distance measurements where greater precision is required. See bytes 69–70 and 71–72 of the trace identification header.

(g) The energy source and geophone group coordinates designated in bytes 73–88 of the trace identification header can be measured in either length or latitude and longitude. The measurement unit used is specified in bytes 89–90 of the trace header. For the latitude/longitude system, the coordinate values are expressed in seconds of arc.

(h) All velocities are in feet per second or meters per second, and these units are not mixed within a reel.

(i) Elevation is represented by "+" above "—" below mean sea level.

(10) The binary coded information convention is defined in Figure 1-C.

DESCRIPTION OF REEL IDENTIFICATION HEADER

The reel identification header (Figure 2) consists of 3600 bytes and is divided into two parts:

(1) The card image EBCDIC block (3200 bytes—40 cards equivalent) followed by an IBG.

(2) The binary coded block (400 bytes) followed by an IBG.

The EBCDIC part of the reel header describes the data from a line of shotpoints in a fixed specified format consisting of 40 card images with each image containing 80 bytes. All unused card image characters are EBCDIC Blank. Card image numbers 23 through 39 are unassigned for optional use. Each card image should contain the character "C" in the first card column. Each 80 bytes would yield one line of format print to produce the form shown in Figure 2-A.

The binary coded section of the reel header consists of 400 bytes of information common to the seismic data on the related reel as shown in Figure 2-B. There are 60 bytes assigned; 340 are unassigned for optional use.

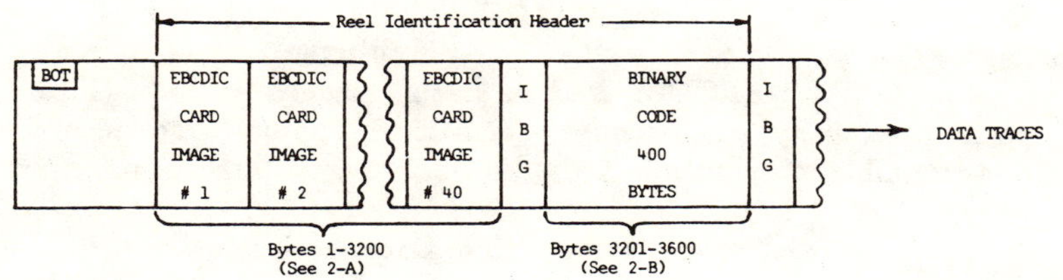

2 - A. EBCDIC CARD IMAGES Free form coding, left justified - 40 card images, 80 bytes per card - card image numbers 23-39 unassigned, for optional information.

CARD IMAGE NUMBER CODE KEY: 0=Zero, Ø=Alpha O, 1=One, I=Alpha I

```
C 1 CLIENT                          COMPANY                       CREW NØ
C 2 LINE                AREA                          MAP ID
C 3 REEL NØ               DAY-START ØF REEL    YEAR        ØBSERVER
C 4 INSTRUMENT: MFG                MØDEL              SERIAL NØ
C 5 DATA TRACES/RECØRD             AUXILIARY TRACES/RECØRD        CDP FØLD
C 6 SAMPLE INTERNAL              SAMPLES/TRACE       BITS/IN        BYTES/SAMPLE
C 7 RECØRDING FØRMAT             FØRMAT THIS REEL        MEASUREMENT SYSTEM
C 8 SAMPLE CØDE: FLØATING PT       FIXED PT       FIXED PT-GAIN      CØRRELATED
C 9 GAIN TYPE: FIXED          BINARY     FLØATING PØINT      ØTHER
C10 FILTERS: ALIAS      HZ NØTCH     HZ BAND     -     HZ SLØPE    -    DB/ØCT
C11 SØURCE: TYPE             NUMBER/PØINT        PØINT INTERVAL
C12     PATTERN:                          LENGTH        WIDTH
C13 SWEEP: START    HZ END      HZ   LENGTH      MS  CHANNEL NØ       TYPE
C14 TAPER: START LENGTH       MS   END LENGTH      MS   TYPE
C15 SPREAD: ØFFSET         MAX DISTANCE         GRØUP INTERVAL
C16 GEØPHØNES: PER GRØUP       SPACING       FREQUENCY     MFG          MØDEL
C17     PATTERN:                          LENGTH        WIDTH
C18 TRACES SØRTED BY: RECØRD        CDP      ØTHER
C19 AMPLITUDE RECØVERY: NØNE       SPHERICAL DIV      AGC      ØTHER
C20 MAP PRØJECTIØN                    ZØNE ID      CØØRDINATE UNITS
C21 PRØCESSING:
C22 PRØCESSING:
C23
C24
C25
C26
C27
C28
C29
C30
C31
C32
C33
C34
C35
C36
C37
C38
C39
C40 END EBCDIC
```

Fig. 2A. Reel identification header. Part 1, the EBCDIC card image block.

There are certain bytes of information that may not apply to a particular recording or processing procedure. It is strongly recommended that bytes designated with an asterisk (*) in Figures 2-B and 3-E always contain the required information.

The data in the reel identification header could be printed and edited prior to the actual input of seismic data for processing. A complete header listing of both the EBCDIC and binary parts would accompany an exchange tape and serve as a table of contents and summary of specifications for that reel of seismic data. No more than one line of seismic data is permitted on any one reel. Additional reels would be used for long lines, and each reel must start with a reel identification header.

DESCRIPTION OF THE TRACE DATA BLOCK

Each trace data block (Figure 3) consists of a fixed 240-byte trace identification header and the seismic trace data. Each trace data block is separated from the next by an IBG. The trace header is written in binary code (refer to Figure 1-C for the binary code information) and is detailed in Figure 3-E.

The trace data samples can be written in

2-B. BINARY CODE-Right justified

Byte Numbers		Description
3201-3204		Job identification number.
3205-3208	*	Line number (only one line per reel).
3209-3212	*	Reel number.
3213-3214	*	Number of data traces per record (includes dummy and zero traces inserted to fill out the record or common depth point).
3215-3216	*	Number of auxiliary traces per record (includes sweep, timing, gain, sync, and all other nondata traces).
3217-3218	*	Sample interval in μsec (for this reel of data). } Designated in microseconds to accommodate sample intervals less than one millisecond.
3219-3220		Sample interval in μsec (for original field recording). } Designated in microseconds to accommodate sample intervals less than one millisecond.
3221-3222	*	Number of samples per data trace (for this reel of data).
3223-3224		Number of samples per data trace (for original field recording).
3225-3226	*	Data sample format code: 1 = floating point (4 bytes); 2 = fixed point (4 bytes); 3 = fixed point (2 bytes); 4 = fixed point w/gain code (4 bytes). Auxiliary traces use the same number of bytes per sample.
3227-3228	*	CDP fold (expected number of data traces per CDP ensemble).
3229-3230		Trace sorting code: 1 = as recorded (no sorting); 2 = CDP ensemble; 3 = single fold continuous profile; 4 = horizontally stacked
3231-3232		Vertical sum code: 1 = no sum, 2 = two sum, ..., N = N sum (N = 32,767)
3233-3234		Sweep frequency at start.
3235-3236		Sweep frequency at end.
3237-3238		Sweep length (msec).
3239-3240		Sweep type code: 1 = linear; 2 = parabolic; 3 = exponential; 4 = other
3241-3242		Trace number of sweep channel.
3243-3244		Sweep trace taper length in msec at start if tapered (the taper starts at zero time and is effective for this length).
3245-3246		Sweep trace taper length in msec at end (the ending taper starts at sweep length minus the taper length at end).
3247-3248		Taper type: 1 = linear; 2 = $\cos^2$; 3 = other
3249-3250		Correlated data traces: 1 = no; 2 = yes
3251-3252		Binary gain recovered: 1 = yes; 2 = no
3253-3254		Amplitude recovery method: 1 = none; 2 = spherical divergence; 3 = AGC; 4 = other
3255-3256	*	Measurement system: 1 = meters; 2 = feet
3257-3258		Impulse signal Polarity: 1 = Increase in pressure or upward geophone case movement gives negative number on tape. 2 = Increase in pressure or upward geophone case movement gives positive number on tape.
3259-3260		Vibratory polarity code: Seismic signal lags pilot signal by: 1 = 337.5° to 22.5°; 2 = 22.5° to 67.5°; 3 = 67.5° to 112.5°; 4 = 112.5° to 157.5°; 5 = 157.5° to 202.5°; 6 = 202.5° to 247.5°; 7 = 247.5° to 292.5°; 8 = 292.5° to 337.5°
3261-3600		Unassigned--for optional information.

*Strongly recommended that this information always be recorded.

FIG. 2B. Reel identification header. Part 2, the binary coded block.

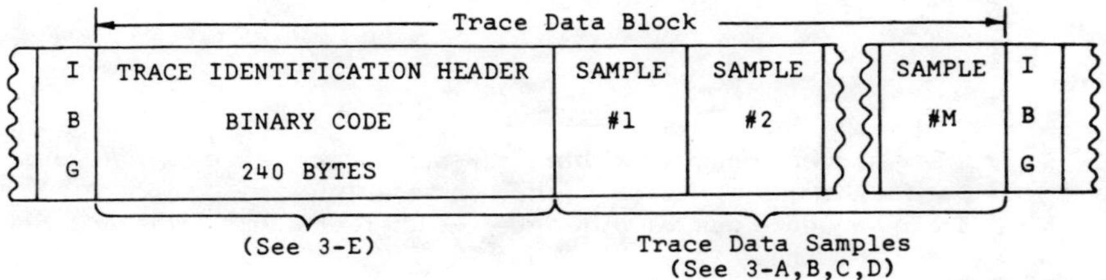

TRACE DATA SAMPLE FORMATS

3-A — 32 Bit Floating Point Format, Sample Code=1

Bit Number	Byte 1	Byte 2	Byte 3	Byte 4
P	P	P	P	P
0	Q_S	Q_F	Q_F	Q_F
1	Q_C	Q_F	Q_F	Q_F
2	Q_C	Q_F	Q_F	Q_F
3	Q_C	Q_F	Q_F	Q_F
4	Q_C	Q_F	Q_F	Q_F
5	Q_C	Q_F	Q_F	Q_F
6	Q_C	Q_F	Q_F	Q_F
7	Q_C	Q_F	Q_F	Q_F

3-B — 32 Bit Fixed Point Format, Sample Code=2

Byte 1	Byte 2	Byte 3	Byte 4
P	P	P	P
Q_S	Q_D	Q_D	Q_D
Q_D	Q_D	Q_D	Q_D
Q_D	Q_D	Q_D	Q_D
Q_D	Q_D	Q_D	Q_D
Q_D	Q_D	Q_D	Q_D
Q_D	Q_D	Q_D	Q_D
Q_D	Q_D	Q_D	Q_D
Q_D	Q_D	Q_D	Q_D

3-C — 16 Bit Fixed Point Format, Sample Code=3

Byte 1	Byte 2
P	P
Q_S	Q_D
Q_D	Q_D
Q_D	Q_D
Q_D	Q_D
Q_D	Q_D
Q_D	Q_D
Q_D	Q_D
Q_D	Q_D

3-D — 32 Bit Fixed Point Format With Gain Values, Sample Code=4

Byte Number: 1	2	3	4
P	P	P	P
0	2^7	Q_S	Q_D
0	2^6	Q_D	Q_D
0	2^5	Q_D	Q_D
0	2^4	Q_D	Q_D
0	2^3	Q_D	Q_D
0	2^2	Q_D	Q_D
0	2^1	Q_D	Q_D
0	2^0	Q_D	Q_D

NOTE: Least significant bit is always in bit position 7 of byte 4 (or byte 2 for 3-C).

Q_S = Sign bit

Q_C = Characteristic

Q_F = Fraction

Q_D = Data bits

FIG. 3A-D. Trace data block. Four data sample options.

one of the four data sample formats described in Figures 3-A, 3-B, 3-C, and 3-D. The trace data format for each reel is identified in bytes 3225–3226 of the reel identification header. Only one data sample format is permitted within each reel.

Figure 3-A details a 32-bit, floating point format in which each data value of a seismic channel is recorded in four successive bytes, in IBM compatible floating point notation as defined in IBM Form GA 22-6821.

The four bytes form a 32-bit word consisting of the sign bit Q_S, a seven-bit characteristic Q_C, and a 24-bit fraction Q_F. Q_S indicates signal polarity and is a one for a negative value. Q_C signifies a power of 16 expressed in excess 64 binary notation allowing both negative and positive powers of 16 to be represented by a

Byte Numbers	Description
1 - 4	* Trace sequence number within line--numbers continue to increase if additional reels are required on same line.
5 - 8	Trace sequence number within reel--each reel starts with trace number one.
9 -12	* Original field record number.
13-16	* Trace number within the original field record.
17-20	Energy source point number--used when more than one record occurs at the same effective surface location.
21-24	CDP ensemble number
25-28	Trace number within the CDP ensemble--each ensemble starts with trace number one.
29-30	* Trace identification code: 1 = seismic data 4 = time break 7 = timing 2 = dead 5 = uphole 8 = water break 3 = dummy 6 = sweep 9----, N = optional use (N = 32,767)
31-32	Number of vertically summed traces yielding this trace. (1 is one trace, 2 is two summed traces, etc.)
33-34	Number of horizontally stacked traces yielding this trace. (1 is one trace, 2 is two stacked traces, etc.)
35-36	Data use: 1 = production. 2 = test.
37-40	Distance from source point to receiver group (negative if opposite to direction in which line is shot).
41-44	Receiver group elevation; all elevations above sea level are positive and below sea level are negative.
45-48	Surface elevation at source.
49-52	Source depth below surface (a positive number).
53-56	Datum elevation at receiver group.
57-60	Datum elevation at source.
61-64	Water depth at source.
65-68	Water depth at group.
69-70	Scaler to be applied to all elevations and depths specified in bytes 41-68 to give the real value. Scaler = 1, ±10, ±100, ±1000, or ±10,000. If positive, scaler is used as a multiplier; if negative, scaler is used as a divisor.
71-72	Scaler to be applied to all coordinates specified in bytes 73-88 to give the real value. Scaler = 1, ±10, ±100, ±1000, or ±10,000. If positive, scaler is used as a multiplier; if negative, scaler is used as divisor.
73-76	Source coordinate - X.
77-80	Source coordinate - Y.
81-84	Group coordinate - X.
85-88	Group coordinate - Y.
	(73-88:) If the coordinate units are in seconds of arc, the X values represent longitude and the Y values latitude. A positive value designates the number of seconds east of Greenwich Meridian or north of the equator and a negative value designates the number of seconds south or west.
89-90	Coordinate units: 1 = length (meters or feet). 2 = seconds of arc.
91-92	Weathering velocity.
93-94	Subweathering velocity.
95-96	Uphole time at source.
97-98	Uphole time at group.

FIG. 3E. Trace identification header written in binary code.

Byte Numbers	Description
99-100	Source static correction.
101-102	Group static correction.
103-104	Total static applied. (Zero if no static has been applied.)
105-106	Lag time A. Time in ms. between end of 240-byte trace identification header and time break. Positive if time break occurs after end of header, negative if time break occurs before end of header. Time break is defined as the initiation pulse which may be recorded on an auxiliary trace or as otherwise specified by the recording system.
107-108	Lag Time B. Time in ms. between time break and the initiation time of the energy source. May be positive or negative.
109-110	Delay recording time. Time in ms. between initiation time of energy source and time when recording of data samples begins. (For deep water work if data recording does not start at zero time.)
111-112	Mute time--start.
113-114	Mute time--end.
115-116	* Number of samples in this trace.
117-118	* Sample interval in μs for this trace.
119-120	Gain type of field instruments: 1 = fixed. 2 = binary. 3 = floating point. 4 --- N = optional use.
121-122	Instrument gain constant.
123-124	Instrument early or initial gain (db).
125-126	Correlated: 1 = no. 2 = yes.
127-128	Sweep frequency at start.
129-130	Sweep frequency at end.
131-132	Sweep length in ms.
133-134	Sweep type: 1 = linear. 2 = parabolic. 3 = exponential. 4 = other.
135-136	Sweep trace taper length at start in ms.
137-138	Sweep trace taper length at end in ms.
139-140	Taper type: 1 = linear. 2 = $\cos^2$. 3 = other.
141-142	Alias filter frequency, if used.
143-144	Alias filter slope
145-146	Notch filter frequency, if used.
147-148	Notch filter slope.
149-150	Low cut frequency, if used.
151-152	High cut frequency, if used.
153-154	Low cut slope
155-156	High cut slope
157-158	Year data recorded.
159-160	Day of year.
161-162	Hour of day (24 hour clock)
163-164	Minute of hour.
165-166	Second of minute.
167-168	Time basis code: 1 = local. 2 = GMT. 3 = other.
169-170	Trace weighting factor--defined as 2^{-N} volts for the least significant bit. (N = 0, 1, ... 32,767.)
171-172	Geophone group number of roll switch position one.
173-174	Geophone group number of trace number one within original field record.
175-176	Geophone group number of last trace within original field record.
177-178	Gap size (total number of groups dropped).
179-180	Overtravel associated with taper at beginning or end of line: 1 = down (or behind). 2 = up (or ahead).
181-240	Unassigned--for optional information.

* Strongly recommended that this information always be recorded.

FIG. 3E. Trace identification header written in binary code (cont.)

true number. Q_F is a six hexadecimal digit (24 binary bit) number with a radix point to the left of the most significant digit. The data value represented by a floating point number is

$$Q_S \times 16^{(Q_C-64)} \times Q_F.$$

Figure 3-B details a 32-bit, fixed point format and each data value of a seismic channel is recorded in four successive bytes. This format consists of a sign bit Q_S (one represents negative) and 31 data bits Q_D with a radix point at the right of the least significant digit.

Figure 3-C represents a 16-bit, fixed point format, and each data value of a seismic channel is recorded in two successive bytes. This format is similar to Figure 3-B except there are 15 data bits Q_D.

Figure 3-D represents a 32-bit, fixed point format with gain values. The first byte of this format is all zeros. The second byte provides eight available gain bits 2^0 through 2^7. The last two bytes are identical to Figure 3-C.

In all four data formats, the channel or trace data should represent the absolute input voltage at the recording instrument. The 32-bit, floating point field format defined as the SEG C (Meiners et al, 1972) comprehends the input voltage level. The fixed point formats 3-B and 3-C require a trace weighting factor (trace identification header, bytes 169–170), defined as 2^{-n} volts for the least significant bit, to comprehend the absolute input voltage level.

In cases where processing parameters such as amplitude recovery are present, the type of amplitude recovery can be described in the appropriate reel identification header sections, and the algorithm described in the unassigned portions.

CONCLUSION

Individual oil companies and contractors may be convinced of their own format's merits, but the use of this recommended exchange demultiplexed format must be given serious consideration in order to achieve some level of industry standardization. Much thought and many suggestions from users have been utilized in establishing a flexible format that yields specifics and can be used by all companies in the industry.

Adoption and use of this format will save substantial sums of money in computer time and programming effort in the future.

ACKNOWLEDGMENTS

Grateful appreciation goes to many companies and individuals for their suggestions at the start of the subcommittee's work and for their final recommendations. We are also thankful for the assistance of Fred Tischler, Texas Instruments, who was the original subcommittee chairman.

REFERENCES

Meiners, E. P., Lenz, L. L., Dalby, A. E., and Hornsby, J. M., 1972, Recommended standards for digital tape formats: Geophysics, v. 37, p. 36–44.

Northwood, E. J., Wisinger, R. C., and Bradley, J. J., 1967, Recommended standards for digital tape formats: Geophysics, v. 32, p. 1073–1084.

Digital field tape format standards—SEG-D

SEG Subcommittee on Digital Tape Formats*

The evolving nature of the seismic data acquisition requires a corresponding change in the recording techniques of these data. Several recording formats have been defined in the past to accommodate the previous changes in instrumentation. They were: SEG's A, B, C, and Y. The following described format includes the best features from these previous formats plus additional characteristics.

By using this new tape format, the data may be multiplexed or demultiplexed, the multiplexed records may be gapped or gapless. The data values are recorded in one of six methods; they are: eight bits or sixteen bits using a quaternary or hexadecimal exponent, twenty bits with a binary exponent, or thirty-two bits with a hexadecimal exponent. The diversity of information to be recorded required a record header that would self-define all the recorded data. Each seismic channel can be individually identified as to its start-stop time, sampling interval, filtering and type of data information recorded.

Included are examples of the defined parameters, example calculations, a glossary, and descriptors for the abbreviations used.

INTRODUCTION

During the past 12 years, a number of recommendations have been published describing proposed tape format standards for both seismic field recording and for seismic trade data (ANSI, 1973a, b, 1976). Many of today's needs, however, were not anticipated in these designs. For example, past formats do not provide for submillisecond sampling intervals, are limited in range and slope of filters, are restricted in the number of recording channels, and make no provisions for either multiple sampling intervals or dynamic changes in parameters such as filters or sampling intervals. This proposal's objective is to overcome these restrictions while at the same time maintaining some compatibility with the presently used formats.

We propose a family of formats all with the same header structure, encompassing both multiplexed and demultiplexed data, with 1, 2, 2½, and 4 byte data words. It is not intended to replace SEG's A, B, C, or Y standard formats. (Northwood et al, 1967; Meiners et al, 1972; Barry et al, 1975). The family is given the designation SEG-D followed by a code number defining the type of data word, and the designation "multiplexed" or "demultiplexed." The code numbers recommended are listed in Appendix A under Bytes 3 and 4 of the general header.

GENERAL FORMAT

The following requirements have been incorporated:

(1) The nine track one-half inch tape is retained, but extension to 6250 BPI is permitted as long as it is compatible with the ANSI standard (ANSI 1973a, b, 1976).

(2) Both multiplexed and demultiplexed field tape formats are represented.

(3) Various word length formats are provided to allow trade-offs to be made when recording large numbers of channels at short sampling intervals.

(4) Multiple sets of data channels can be recorded, each operating at a different sam-

*D.A. Cavers (Chairman), Gulf Research and Development, Box 36506, Houston, TX 77036; P.E. Carroll, Texas Instruments, P.O. Box 1444, M.S. 617, Houston, TX 77001; E.P. Meiners, Chevron Geophysical, Box 36487, Houston, TX 77036; C.W. Racer, Chevron Geophysical, 8114 Pella, Houston, TX 77036; L.E. Siems, Litton Resources, 3930 Westholme Drive, Houston, TX 77063; M.G. Sojourner, Input/Output, 8009 Harwin Drive, Houston, TX 77036; J.L. Twombly, GUS Manufacturing, 10486 Dauwood, El Paso, TX 79925; J.A. Weigand, Texaco Inc., 4800 Fournace Place E302, Bellaire, TX 77401.

pling interval, or each having other parameter differences (e.g., filters) during the record. The parameters can be constant throughout the record, or can be dynamically switched at some predetermined time.

(5) The format is self-defining. Information in the header specifies the length of the header and the length of the data record.

(6) A one's complement number system is selected for the fractional part in all binary and quaternary exponent data recording methods. A sign and magnitude number system is selected for the fractional part of all hexadecimal exponent data recording methods.

(7) A unique four byte start of scan code is used. Status bits are provided to indicate dynamic changes starting with this scan.

(8) Submillisecond timing words are incorporated to allow shorter sampling interval systems to have a unique timing word in each scan.

(9) One start of scan code followed by a timing word is utilized in all multiplexed formats in order to facilitate software and hardware synchronization.

(10) Provisions are made for recording the instrument sample skew in the header. (This does not include cable propagation delay which must be established by the user.)

FORMAT OVERVIEW

The overall tape layout is as follows:

Leader
BOT
Seismic record file 1
EOF
Seismic record file 2
EOF
.
.
.
.
Seismic record file *n*
EOF
EOF
EOT
Trailer

As shown in Figure 1, each seismic record consists of a file of two or more blocks. The first block contains the record headers. Following the header block will be one or more blocks of multiplexed data (Figure 2) or of demultiplexed data (Figure 3). Seismic records are separated by one end of file mark (EOF). Two EOF marks indicate that there are no more seismic records on the tape.

HEADER BLOCK—FUNCTIONAL DESCRIPTION

The record header block is a single block of information separated from the data by a standard interblock gap. The record header block is composed of a general header, one or more scan type headers and, optionally, the extended and external headers (Figure 1). All header information is packed BCD unless otherwise specified, with bit positions 0–3 being the most significant digit. Appendix A has a detailed listing of the information contained in the general and scan type headers.

General header (required)

The general header is 32 bytes long and contains information similar to SEG A, B, and C headers (Figure 4). Abbreviations (Appendix D) are as close as possible to those used in previous formats.

Scan type header (required)

The scan type header is new. It is used to describe the information of the recorded channels [filters, sampling intervals, sample skew, etc. (Figure 6)]. The scan type header is composed of one or more channel set descriptors followed by skew information. The channel set descriptors must appear in the same order as their respective channel sets will appear within a base scan interval. A channel set, which is part of a scan type, is defined as a group of channels all recorded with identical recording parameters. One or more channel sets can be recorded concurrently within one scan type. In addition, there can be multiple scan types to permit dynamic scan type changes during the record (e.g., 12 channels at ½ msec switched at about 1 sec to 48 channels at 2 msec). Where there are dynamic changes, scan type header 1 describes the first part of the record, scan type header 2 the second part, etc. Within the scan type header, each channel set descriptor is composed of a 32 byte field (Figure 5), and up to 99 channel set descriptors may be present. In addition, up to 99 scan type headers may be utilized in a record.

Following the channel set descriptors of a scan type are a number of 32 byte fields (SK, specified in Byte 30 of the general header) that specify sample skew. Sample skew (SS) is recorded in a single byte for each sample of each subscan of each channel set, in the same order as the samples are recorded in the scan. Each byte represents a fractional part of the base scan interval (Byte 23 of the general header). The resolution is 1/256 of this interval. For instance, if the base scan interval is 2 msec, the least significant bit in the sample skew byte is 1/256 of 2 msec or 7.8125 μsec. The reference point for the skew is the timing word (T) such that the timing word represents zero skew. The actual time the sample was taken is obtained by adding the timing word (T) to the product of the sample skew (SS) and the base scan interval. (Actual time = $T + SS \times I$.)

The following is a list of ground rules for the scan type header:

(1) The order in which channel sets are described in the header will be the same as the order in which the data are recorded for each channel set. (See multiplexed data example 6.)

(2) In a scan type header containing multiple channel set descriptors with different sampling intervals, each channel set descriptor will appear only once in each scan type header. Within the data block, however, shorter sampling interval data are recorded more frequently (i.e., within a base scan interval, channel sets of a shorter sampling interval will appear as multiple subscans). The number of subscans of a channel set is equal to the quotient of the base scan interval divided by the channel set sampling interval [see Byte 12 of the descriptor for this channel set (Figure 5)].

(3) In the case of multiple scan type records, such as the dynamically switched sampling interval case, each scan type will contain the same number of channel sets. Any unused channel sets needed in a scan type must be so indicated by setting Bytes 9 and 10 (channels per channel set) to zero in the channel set descriptor (Appendix A).

(4) In multiple scan type records, the number of bytes per base scan interval must remain a constant for all scan types recorded.

(5) The data recording method is not permitted to change on a reel of tape. However, multiplexed and demultiplexed data may appear on the same reel. For example, one or more multiplexed or demultiplexed records may be recorded on tape followed by a multiplexed or demultiplexed stacked record of the same data with the same word length and data recording method.

(6) Although not essential, it is suggested that the channel set order within a scan type be: auxiliary channels, long sampling interval channels, short sampling interval channels. All channel sets of the same sampling interval should be contiguous (see Example 4).

Extended header (optional)

The extended header provides additional areas to be used by equipment manufacturers to interface directly with their equipment. An example of this would be a vertical stacking unit used in conjunction with the data acquisition system (e.g., records per stack, records rejected in stack, type of stack, etc.). Since the nature of this data will depend heavily on the equipment and processes being applied, it will be the responsibility of the equipment manufacturer to establish a format and document this area. Byte 31 of the general header contains the number of 32 byte fields in the extended header.

External header (optional)

The external header provides a means of recording special user desired information in the header block. Some examples of this are roll box information, crew data, survey information, and a multiplicity of marine parameters. This data format will be defined and documented by the end user. The means of putting this information into the header has usually been provided by the equipment manufacturer. Byte 32 of the general header contains the number of 32 byte fields in the external header.

DATA BODY

Multiplexed

The multiplexed data may be gapped or gapless. A gapless multiplexed data block (Figure 2) is one continuous block of data separated from the header block by an interblock gap, and divided into an integral number of scans as defined in the header. If the multiplexed data body is gapped, each block must begin with a start-of-scan and must contain an integral number of scans. Bytes 24 and 25 of the general header indicate the number of scans in a block. Zero indicates gapless data.

In each scan the first eight bytes are dedicated to the start-of-scan code and timing word as shown in Figure 8. The sync code and timing word provide a means of recovering data that might otherwise be lost, and allow various checks to be made on the integrity of the data while it is being read during processing.

The 8 byte start-of-scan code and timing word overhead must be considered (counted) when computing the number of bytes per scan. The 8 byte start-of-scan and timing word format will remain the same in each of the different data recording methods. The start-of-scan code in each of the data recording methods described here must be maintained as a unique 4 byte code; therefore, there will be certain restrictions on the data in each of the different data recording methods which are covered in the individual format descriptions. (See general header Bytes 3 and 4.)

The 4 byte (start of scan) code as shown in Figure 8 and below is composed as follows: The first three bytes are all one's, and Bits 6 and 7 of Byte 4 must be a zero and a one, respectively, to guarantee uniqueness of the start of scan. The remaining bits (0–5) of Byte 4 are undefined (*x*) or are used as follows:

Start of scan

Bit	0	1	2	3	4	5	6	7
Byte 1	1	1	1	1	1	1	1	1
Byte 2	1	1	1	1	1	1	1	1
Byte 3	1	1	1	1	1	1	1	1
Byte 4	x	TWI	ITB	DP	x	x	0	1

TWI.—This bit is included as an integrity check on time break. It changes from a zero to a one at the close of the time break window. Random variations in the time of this change indicate a problem in the fire control system. The presence of a one in the TWI bit of the first start of scan of a record indicates that time break was not detected and recording commenced at the end of the time break window.

ITB.—Internal time break is recorded as a 1 for the entire record if an abnormal condition is detected in the synchronization of the system timing with the energy source timing. Otherwise it is recorded as a zero.

DP.—The dynamic parameter change bit is recorded as a zero for the first scan of every record and remains a zero throughout the first scan type. It is switched from a zero to a one to indicate the first scan of the second scan type which contains the first data taken after a dynamic parameter change has occurred. It will remain a one throughout the second scan type until another dynamic parameter change occurs, at which time it will be switched to a zero. It is alternately switched at each subsequent scan type. For records without dynamic parameter changes, it is recorded as a zero throughout the record.

The next three bytes (Bytes 5–7 of the scan) are dedicated to a binary timing word as shown in Figure 8 and below. Byte 8 is written as all zeros. The timing word is in milliseconds and has the following bit weight assignments:

Timing word

Bit	0	1	2	3	4	5	6	7
Byte 5	2^{15}	2^{14}	2^{13}	2^{12}	2^{11}	2^{10}	2^{9}	2^{8}
Byte 6	2^{7}	2^{6}	2^{5}	2^{4}	2^{3}	2^{2}	2^{1}	2^{0}
Byte 7	2^{-1}	2^{-2}	2^{-3}	2^{-4}	2^{-5}	2^{-6}	2^{-7}	2^{-8}
Byte 8	0	0	0	0	0	0	0	0

The timing word LSB (2^{-8}) is equal to 1/256 msec, and the MSB (2^{15}) is equal to 32,768 msec. The timing word for each scan is equal to the elapsed time from zero time to the start of that scan. Timing words of from 0 to 65,535.9961 msec are codable. For longer recordings the timing word may overflow to zero and then continue.

The first scan of data has typically started with timing word zero. However, this is not a requirement. In a sampling system, it is not always practical to resynchronize the system even though most seismic data acquisition systems have to date. Possible reasons for not wanting to resynchronize could be digital filtering, communication restrictions, etc.

Whether the system is resynchronized or not, the timing word will contain the time from the energy source event to the start of scan of interest. For example, assume the sampling interval is 2 msec, the system does not resynchronize, and the energy source event occurs 1 + 9/256 msec before the next normal start of scan. The timing word values would be:

First timing word	0 + 1 + 9/256 msec
Second	2 + 1 + 9/256 msec
Third	4 + 1 + 9/256 msec
Fourth	6 + 1 + 9/256 msec
⋮	
One-thousandth timing word	1998 + 1 + 9/256 msec

Demultiplexed

The demultiplexed data format (Figure 3) is simply an extension of the multiplexed format previously described (Figure 2). In an equivalent system, it would utilize the same header block as its multiplexed counterpart with the exception of the format code in Bytes 3 and 4 and bytes per scan in Bytes 20, 21, and 22 of the general header. The data body portion of the multiplexed format (as shown in Figure 2) is replaced with individual trace blocks (Figure 3) including a trace header (see Figure 9 and below) for all channels in each channel set of each scan type. Each trace block is separated by a standard inter block gap of erased tape and is a sequential set of points from one channel in one channel set. The data recording methods utilized within the trace blocks are described later in the section titled "Data recording method."

Trace header.—The trace header length is 20 bytes and is an identifier that precedes each channel's data. The trace header and the trace data are recorded as one block of data. A trace is restricted to one channel of data from one channel set of one scan type. All the information in the trace header is taken directly from the general header and scan type header with the exception of the channel number and the time break window end time. File number, scan type, channel set, and channel (or trace) number are recorded in packed BCD.

it represents the time from time break to the end of the time break window. Random variations in this time indicate a problem in the fire control system. The presence of a value less than the base scan interval indicates that time break was not detected and recording commenced at the end of the time break window.

DATA RECORDING METHOD

To accommodate diverse recording needs, the data recording utilizes sample sizes of 8, 16, 20, and 32 bits.

The data word is a number representation of the sign and magnitude of the instantaneous voltage presented to the system. It is not an indication of how the hardware gain system functions. The output of stepped gain systems may be represented as a binary mantissa and a binary exponent of base 2, 4, or 16 (binary, quaternary, or hexadecimal system).

Trace header

	Bit	0	1	2	3	4	5	6	7
File number	Byte 1	F_1	F_1	F_1	F_1	F_2	F_2	F_2	F_2
	Byte 2	F_3	F_3	F_3	F_3	F_4	F_4	F_4	F_4
Scan type	Byte 3	ST_1	ST_1	ST_1	ST_1	ST_2	ST_2	ST_2	ST_2
Channel set	Byte 4	CN_1	CN_1	CN_1	CN_1	CN_2	CN_2	CN_2	CN_2
Channel or	Byte 5	TN_1	TN_1	TN_1	TN_1	TN_2	TN_2	TN_2	TN_2
trace number	Byte 6	TN_3	TN_3	TN_3	TN_3	TN_4	TN_4	TN_4	TN_4
First timing	Byte 7	T_{15}	T_{14}	T_{13}	T_{12}	T_{11}	T_{10}	T_9	T_8
word	Byte 8	T_7	T_6	T_5	T_4	T_3	T_2	T_1	T_0
	Byte 9	T_{-1}	T_{-2}	T_{-3}	T_{-4}	T_{-5}	T_{-6}	T_{-7}	T_{-8}
	Byte 10	0	0	0	0	0	0	0	0
Sample skew	Byte 11	SS_{-1}	SS_{-2}	SS_{-3}	SS_{-4}	SS_{-5}	SS_{-6}	SS_{-7}	SS_{-8}
	Byte 12	0	0	0	0	0	0	0	0
Time break	Byte 13	TW_{15}	TW_{14}	TW_{13}	TW_{12}	TW_{11}	TW_{10}	TW_9	TW_8
window end	Byte 14	TW_7	TW_6	TW_5	TW_4	TW_3	TW_2	TW_1	TW_0
	Byte 15	TW_{-1}	TW_{-2}	t $_{,-3}$	TW_{-4}	TW_{-5}	TW_{-6}	TW_{-7}	TW_{-8}
	Byte 16	0	0	0	0	0	0	0	0
		⋮	⋮	⋮	⋮	⋮	⋮	⋮	⋮
	Byte 20	0	0	0	0	0	0	0	0

Bytes 7, 8, and 9 comprise the timing word that would accompany the first sample if these data were written in the multiplexed format. To obtain the exact sample time, the actual sample skew time (Byte 11 multiplied by the base scan interval) must be added to the time recorded in Bytes 7, 8, and 9.

Byte 11 contains sample skew of the first sample of this trace. This is identical to the first byte of sample skew for this channel in the scan type header.

Bytes 13, 14, and 15 are included as an integrity check on time break. They comprise the timing word of the scan in which TWI changed to a one. Thus,

Following are descriptions of each of the data recording methods permitted. The same number system is to be used on all samples in a record, including auxiliary and all other types of channels. All recording methods are valid for multiplexed and demultiplexed records. The 2½ byte binary demultiplexed method uses the LSB whereas the comparable multiplexed method does not (in order to preserve the uniqueness of the start of scan code).

1 byte quaternary exponent data recording method

The following illustrates the 8 bit word and the corresponding bit weights:

Bit	0	1	2	3	4	5	6	7
Byte 1	S	C_2	C_1	C_0	Q_{-1}	Q_{-2}	Q_{-3}	Q_{-4}

S=sign bit.—(One = negative number).
C=quaternary exponent.—This is a three bit positive binary exponent of 4 written as 4^{CCC} where CCC can assume values from 0–7.
Q_{1-4}-fraction.—This is a 4 bit one's complement binary fraction. The radix point is to the left of the most significant bit (Q_{-1}) with the MSB being defined as 2^{-1}. The fraction can have values from $-1 + 2^{-4}$ to $1 - 2^{-4}$. In order to guarantee the uniqueness of the start of scan, negative zero is invalid and must be converted to positive zero.
Input signal = S.QQQQ $\times 4^{CCC} \times 2^{MP}$ millivolts where 2^{MP} is the value required to descale the data sample to the recording system input level. MP is defined in Byte 8 of each channel set descriptor in the scan type header.

2 byte quaternary exponent data recording method

The following illustrated the 16-bit word and the corresponding bit weights:

Bit	0	1	2	3	4	5	6	7
Byte 1	S	C_2	C_1	C_0	Q_{-1}	Q_{-2}	Q_{-3}	Q_{-4}
Byte 2	Q_{-5}	Q_{-6}	Q_{-7}	Q_{-8}	Q_{-9}	Q_{-10}	Q_{-11}	Q_{-12}

S=sign bit.—(One = negative number).
C=quaternary exponent.—This is a three bit positive binary exponent of 4 written as 4^{CCC} where CCC can assume values from 0–7.
Q_{1-12} – fraction.—This is a 12 bit one's complement binary fraction. The radix point is to the left of the most significant bit (Q_{-1}) with the MSB being defined as 2^{-1}. The fraction can have values from $-1 + 2^{-12}$ to $1 - 2^{-12}$. In order to guarantee the uniqueness of the start of scan, negative zero is invalid and must be converted to positive zero.
Input signal = S.QQQQ,QQQQ,QQQQ $\times 4^{CCC} \times 2^{MP}$ millivolts where 2^{MP} is the value required to descale the data sample to the recording system input level. MP is defined in Byte 8 of each channel set descriptor in the scan type header.

2½ byte binary exponent data recording method—multiplexed

The following illustrates the 20-bit word and the corresponding bit weights:

Bit	0	1	2	3	4	5	6	7	
Byte 1	C_3	C_2	C_1	C_0	C_3	C_2	C_1	C_0	Exponent for channels 1 thru 4[a]
Byte 2	C_3	C_2	C_1	C_0	C_3	C_2	C_1	C_0	
Byte 3	S	Q_{-1}	Q_{-2}	Q_{-3}	Q_{-4}	Q_{-5}	Q_{-6}	Q_{-7}	Channel 1
Byte 4	Q_{-8}	Q_{-9}	Q_{-10}	Q_{-11}	Q_{-12}	Q_{-13}	Q_{-14}	0	
Byte 5	S	Q_{-1}	Q_{-2}	Q_{-3}	Q_{-4}	Q_{-5}	Q_{-6}	Q_7	Channel 2
Byte 6	Q_{-8}	Q_{-9}	Q_{-10}	Q_{-11}	Q_{-12}	Q_{-13}	Q_{-14}	0	
Byte 7	S	Q_{-1}	Q_{-2}	Q_{-3}	Q_{-4}	Q_{-5}	Q_{-6}	Q_{-7}	Channel 3
Byte 8	Q_{-8}	Q_{-9}	Q_{-10}	Q_{-11}	Q_{-12}	Q_{-13}	Q_{-14}	0	
Byte 9	S	Q_{-1}	Q_{-2}	Q_{-3}	Q_{-4}	Q_{-5}	Q_{-6}	Q_{-7}	Channel 4
Byte 10	Q_{-8}	Q_{-9}	Q_{-10}	Q_{-11}	Q_{-12}	Q_{-13}	Q_{-14}	0	

[a] In the multiplexed format, Bytes 1 and 2 contain the exponents for the following four channels of the scan. The channel numbers are relative and are only to denote position in the four channel subset.

S=sign bit.—(One = negative number).
C=binary exponents.—This is a 4 bit positive binary exponent of 2 written as 2^{CCCC} where CCCC can assume values of 0–15. The four exponents are in channel number order for the four channels starting with channel one in bits 0–3 of Byte 1.
Q^{1-14}-fraction.—This is a 14 bit one's complement binary fraction. The radix point is to the left of the most significant bit (Q_{-1}) with the MSB being defined as 2^{-1}. The sign and fraction can assume values from $1 - 2^{-14}$ to $-1 + 2^{-14}$. Note that bit 7 of the second byte of each sample must be zero in order to guarantee the uniqueness of the start of scan. Negative zero is invalid and must be converted to positive zero.
Input signal = S.QQQQ,QQQQ,QQQQ,QQ $\times$ $2^{CCCC} \times 2^{MP}$ millivolts where 2^{MP} is the value required to descale the data word to the recording system input level. MP is defined in Byte 8 of each of the corresponding channel set descriptors in the scan type header.

Note that in utilizing this data recording method, the number of data channels per channel set must be exactly divisible by 4 in order to preserve the data grouping of this method.

2½ byte binary exponent data recording method—demultiplexed

The following illustrates the 20 bit word and the corresponding bit weights:

Bit	0	1	2	3	4	5	6	7	
Byte 1	C_3	C_2	C_1	C_0	C_3	C_2	C_1	C_0	Exponent for samples 1 thru 4[b]
Byte 2	C_3	C_2	C_1	C_0	C_3	C_2	C_1	C_0	
Byte 3	S	Q_{-1}	Q_{-2}	Q_{-3}	Q_{-4}	Q_{-5}	Q_{-6}	Q_{-7}	Sample 1
Byte 4	Q_{-8}	Q_{-9}	Q_{-10}	Q_{-11}	Q_{-12}	Q_{-13}	Q_{-14}	Q_{-15}	
Byte 5	S	Q_{-1}	Q_{-2}	Q_{-3}	Q_{-4}	Q_{-5}	Q_{-6}	Q_{-7}	Sample 2
Byte 6	Q_{-8}	Q_{-9}	Q_{-10}	Q_{-11}	Q_{-12}	Q_{-13}	Q_{-14}	Q_{-15}	
Byte 7	S	Q_{-1}	Q_{-2}	Q_{-3}	Q_{-4}	Q_{-5}	Q_{-6}	Q_{-7}	Sample 3
Byte 8	Q_{-8}	Q_{-9}	Q_{-10}	Q_{-11}	Q_{-12}	Q_{-13}	Q_{-14}	Q_{-15}	
Byte 9	S	Q_{-1}	Q_{-2}	Q_{-3}	Q_{-4}	Q_{-5}	Q_{-6}	Q_{-7}	Sample 4
Byte 10	Q_{-8}	Q_{-9}	Q_{-10}	Q_{-11}	Q_{-12}	Q_{-13}	Q_{-14}	Q_{-15}	

[b] In the demultiplexed format, Bytes 1 and 2 contain the exponents for the following four samples of the channel. The sample numbers are relative and are only to denote position in the four sample subset.

S=sign bit—(One = negative number).
C=binary exponent.—This is a 4 bit positive binary exponent of 2 written as 2^{CCCC} where CCCC can assume values of 0–15. The four exponents are in sample order for the four samples starting with the first sample in bits 0–3 of Byte 1.
Q_{1-15}-fraction.—This is a 15 bit one's complement binary fraction. The radix point is to the left of the most significant bit (Q_{-1}) with the MSB being defined as 2^{-1}. The sign and fraction can assume values from $1 - 2^{-15}$ to $-1 + 2^{-15}$. Negative zero is invalid and must be converted to positive zero.
Input signal=S.Q, QQQ, QQQQ, QQQQ, QQQ $2^{CCCC} \times 2^{MP}$ millivolts where 2^{MP} is the value required to descale the data word to the recording system input level. MP is defined in Byte 8 of each of the corresponding channel set descriptors in the scan type header.

Note that in utilizing this data recording method, the number of samples per channel must be exactly divisible by 4 in order to preserve the data grouping of this method.

---

1 byte hexadecimal exponent data recording method

The following illustrates the 8-bit word and the corresponding bit weights:

Bit	0	1	2	3	4	5	6	7
Byte 1	S	C_1	C_0	Q_{-1}	Q_{-2}	Q_{-3}	Q_{-4}	Q_{-5}

S=sign bit.—(One = negative number).
C=hexadecimal exponent.—This is a two bit positive binary exponent of 16 written as 16^{CC} where CC can assume values from 0–3.
$Q_{1\text{-}5}$*-fraction.*—This is a 5 bit positive binary fraction. The radix point is to the left of the most significant bit (Q_{-1}) with the MSB being defined as 2^{-1}. The sign and fraction can have any value from $-1 + 2^{-5}$ to $1 - 2^{-5}$. In order to guarantee the uniqueness of the start of scan, an all one's representation (sign = negative, exponent = 3, and fraction = $1 - 2^{-5}$) is invalid. Thus the full range of values allowed is $-(1 - 2^{-4}) \times 16^3$ to $+(1 - 2^{-5}) \times 16^3$.
Input signal = S.QQQQ,Q × $16^{CC} \times 2^{MP}$ millivolts where 2^{MP} is the value required to descale the data sample to the recording system input level. MP is defined in Byte 8 of each channel set descriptor in the scan type header.

---

2 byte hexadecimal exponent data recording method

The following illustrates the 16-bit word and the corresponding bit weights:

Bit	0	1	2	3	4	5	6	7
Byte 1	S	C_1	C_0	Q_{-1}	Q_{-2}	Q_{-3}	Q_{-4}	Q_{-5}
Byte 2	Q_{-6}	Q_{-7}	Q_{-8}	Q_{-9}	Q_{-10}	Q_{-11}	Q_{-12}	Q_{-13}

S=sign bit.—(One = negative number).
C=hexadecimal exponent.—This is a two bit positive binary exponent of 16 written as 16^{CC} where CC can assume values from 0–3.
$Q_{1\text{-}13}$*-fraction.*—This is a 13 bit positive binary fraction. The radix point is to the left of the most significant bit (Q_{-1}) with the MSB being defined as 2^{-1}. The sign and fraction can have any value from $-1 + 2^{-13}$ to $1 - 2^{-13}$. In order to guarantee the uniqueness of the start of scan, an all one's representation (sign = negative, exponent = 3, and fraction = $1 - 2^{-13}$) is invalid. Thus the full range of values allowed is $-(1 - 2^{-12}) \times 16^3$ to $+(1 - 2^{-13}) \times 16^3$.
Input signal = S.QQQQ,QQQQ,QQQQ,Q × $16^{CC} \times 2^{MP}$ millivolts where 2^{MP} is the value required to descale the data sample to the recording system input level. MP is defined in Byte 8 of each channel set descriptor in the scan type header.

---

4 byte hexadecimal exponent data recording method

The following illustrates the 32-bit word and the corresponding bit weights:

Bit	0	1	2	3	4	5	6	7
Byte 1	S	C_6	C_5	C_4	C_3	C_2	C_1	C_0
Byte 2	Q_{-1}	Q_{-2}	Q_{-3}	Q_{-4}	Q_{-5}	Q_{-6}	Q_{-7}	Q_{-8}
Byte 3	Q_{-9}	Q_{-10}	Q_{-11}	Q_{-12}	Q_{-13}	Q_{-14}	Q_{-15}	Q_{-16}
Byte 4	Q_{-17}	Q_{-18}	Q_{-19}	Q_{-20}	Q_{-21}	Q_{-22}	Q_{-23}	0

S=sign bit.—(One = negative number).
C=excess 64 hexadecimal exponent.—This is a binary exponent of 16. It has been biased by 64 such that it represents $16^{(CCCCCCC-64)}$ where CCCCCCC can assume values from 0 to 127.
$Q_{1\text{-}23}$*-magnitude fraction.*—This is a 23 bit positive binary fraction (i.e., the number system is sign and magnitude). The radix point is to the left of the most significant bit (Q_{-1}) with the MSB being defined as 2^{-1}. The sign and fraction can assume values from ($1 - 2^{-23}$ to $-1 + 2^{-23}$). It must always be written as a hexadecimal left justified number. If this fraction is zero, the sign and exponent must also be zero (i.e., the entire word is zero). Note that bit 7 of Byte 4 must be zero in order to guarantee the uniqueness of the start of scan.
Input signal = S. QQQQ, QQQQ, QQQQ, QQQQ, QQQQ, QQQ × $16^{(CCCCCCC-64)} \times 2^{MP}$ millivolts where 2^{MP} is the value required to descale the data sample to the recording system input level. MP is

defined in Byte 8 of each channel set descriptor in the scan type header. This data recording method has more than sufficient range to handle the dynamic range of a typical seismic system. Thus, MP may not be needed to account for any scaling and may be recorded as zero.

EXAMPLES

Three sets of examples are given. The header block set illustrates the header lengths for various combinations of scan types and channel sets. The multiplexed data set of examples illustrates the organization of data within scans for an example which includes multiple scan types and multiple channel sets. The demultiplexed data set of examples illustrates the order of trace blocks for most of the examples.

Header block

The following examples describe the general and scan type headers required for each case. The optional extended and external headers are not included.

Example 1—Typical 24 channel system (Figure 6A)

4 auxiliary channels at 2-msec sampling intervals
24 seismic channels at 2-msec sampling intervals

This system would contain one scan type header because there are no parameter changes during the record. This one scan type header would contain two channel set descriptors, one for the 4 auxiliary channels and one for the 24 seismic channels. The header block in this case would be as follows, assuming no extensions for the extended and external headers:

32 bytes	General header
	Scan type header 1
32 bytes	auxiliary channel set
32 bytes	seismic channel set
32 bytes	sample skew (28 used)
128 bytes	total header block

Example 2—24 channel system with parameter differences (Figure 6B)

4 auxiliary channels
12 seismic channels with 18-Hz low-cut filters
12 seismic channels with 36-Hz low-cut filters
(2-msec sampling on all channels)

This example is similar to the previous one with the exception of needing one additional channel set descriptor, which is required because some of the seismic channels are operating at a different filter setting. The header block in this case would contain one additional 32 byte field.

32 bytes	General header
	Scan type header 1
32 bytes	auxiliary channel set
32 bytes	seismic channel set (18-Hz low cut)
32 bytes	seismic channel set (36-Hz low cut)
32 bytes	sample skew (28 used)
160 bytes	Total header block

Example 3—Large system (Figure 6C)

4 auxiliary channels
240 seismic channels
(4 msec sampling on all channels)

The header block in this case will be longer than in Example 1 because it will have additional bytes of sample skew.

32 bytes	General header
	Scan type header 1
32 bytes	auxiliary channel set
32 bytes	seismic channel set
256 bytes	sample skew (244 used)
352 bytes	Total header block

Example 4—Dual sampling intervals (Figure 7A)

4 auxiliary channels at 2-msec sampling interval
48 seismic channels at 2-msec sampling interval
12 seismic channels at ½-msec sampling interval
(sampled four times in each scan)

In this case there would be three channel set descriptors required to make up the scan type header. The additional channel set descriptor is required to identify the channel set operating at ½ msec.

32 bytes	General header
	Scan type header 1
32 bytes	auxiliary channel set
32 bytes	seismic channel set at 2 msec
32 bytes	seismic channel set at ½ msec
128 bytes	sample skew (100 used)
256 bytes	Total header block

Example 5—Dynamically switched sampling interval (Figure 7B)

Scan type header 1 4 auxiliary channels at 2-msec sampling interval
12 seismic channels at ½-msec sampling interval

Scan type header 2 4 auxiliary channels at 2 msec
48 seismic channels at 2 msec

This is a switched sampling interval system in which 12 of the channels will be sampled for a portion of the record at a ½-msec interval, then it will switch to 48 seismic channels at a 2-msec interval for the remainder of the record.

This system has a parameter change in mid-record and therefore requires a second scan type header to define it. Its header block would be constructed as follows:

32 bytes	General header
	Scan type header 1
32 bytes	4 auxiliary channels at 2 msec
32 bytes	12 seismic channels at ½ msec
64 bytes	Sample skew (52 used)
	Scan type header 2
32 bytes	4 auxiliary channels at 2 msec
32 bytes	48 seismic channels at 2 msec
64 bytes	Sample skew (52 used)
288 bytes	Total header block

Example 6—Dynamically switched sampling interval (Figure 7C)

This, as in Example 5, is a 2-msec base scan interval system that is dynamically switching sampling interval in mid-record. It starts as a 12 channel ½-msec sampling interval system and ends as a 48 channel 2-msec sampling interval system with a common filter. The 12 channels in this example, though, are divided into 2 channel sets, with 6 channels utilizing low-cut filters and the remaining 6 without low-cut filters. The header block would be constructed in the following manner:

32 bytes	General header
	Scan type header 1
32 bytes	4 auxiliary channels at 2 msec
32 bytes	6 seismic channels with low-cut filters at ½ msec
32 bytes	6 seismic channels without low-cut filters at ½ msec
64 bytes	Sample skew (52 used)
	Scan type header 2
32 bytes	4 auxiliary channels at 2 msec
32 bytes	48 seismic channels at 2 msec
32 bytes	0 channels (dummy)
64 bytes	Sample skew (52 used)
352 bytes	Total header block

The dummy channel set descriptor is required in scan type header 2 in order to maintain its length equal to that of scan type header 1. This preserves easy expandability of the format and its self-defining header length capability (see Appendix E4).

Multiplexed data

Example 6.—The multiplexed data block section of the seismic record is organized in the same order as the scan type headers and channel set descriptors they contain. Within each base scan in scan type 1, all of the data for channel set 1 are recorded before channel set 2. Likewise, all of the data for channel set 2 are recorded before channel set 3. Because the sampling interval for channel sets 2 and 3 is one-fourth of the base scan interval, each base scan will contain four subscans (a subscan contains one sample from each channel) of channel sets 2 and 3. Furthermore, all four subscans of channel set 2 are recorded before the four subscans of channel set 3. Thus the base scan interval is 2 msec and the data would be recorded in the following order: (1) start of scan and timing word; (2) four auxiliary channels in channel number sequence; (3a) six seismic channels with low-cut filters in channel number sequence; (3b) the second, third, and fourth set of samples of the channels in 3a; (4a) six seismic channels without low-cut filters in channel number sequence; and (4b) the second, third, and fourth sets of samples of the channels in (4a).

The pictorial representation of the multiplexed format of Example 6 as recorded on tape is shown below. Scan type 1 is:

Start of scan and timing word 2 msec base scan	4 Auxiliary channels sampled once each	4 Subscans of 6 seismic channels with low-cut filters sampled at ½ msec	4 Subscans of 6 seismic channels without low-cut filters sampled at ½ msec

Note that this scan is repeated for the required number of scans until the channel set length is satisfied and the sampling interval is changed.
Scan type 2 is:

Start of scan and timing word	4 Auxiliary channels sampled once each at 2 msec	48 Seismic channels sampled once each at 2 msec

Note that this scan type follows the recording of all of scan type 1 and repeats for the required number of scans. Scan types 1 and 2 have the same number of bytes per scan.

Demultiplexed data

The following examples are the demultiplexed versions of the previous examples given.

Example 1.—Typical 24 channel system

The trace blocks in this case would be in the following order.

4 traces of auxiliary channels
24 traces of seismic data
28 trace blocks total

The channel order above is the same order in which the channel descriptions appear in the scan type header. All other examples of one scan type system follow this pattern.

Example 2.—24 channel system with parameter differences

The trace blocks in this case would be in the following order.

4 traces of auxiliary channels
12 traces of seismic data with 18-Hz low-cut filters
12 traces of seismic data with 36-Hz low-cut filters
28 trace blocks total

Example 3.—Is similar to Example 1, but it has 240 traces of seismic data, resulting in 244 trace blocks total.

Example 4.—Dual sampling intervals

The trace blocks in this case would be in the following order.

4 traces of auxiliary channels
48 traces of seismic data recorded at a 2-msec sampling interval
12 traces of seismic data recorded at a ½-msec sampling interval
64 trace blocks total

Example 5.—Dynamically switched sampling interval

The trace blocks appear in the following order.

4 traces of auxiliary channels, recorded for the length of scan type 1
12 traces of seismic channels, recorded for the length of scan type 1
4 traces of auxiliary channels, recorded for the length of scan type 2
48 traces of seismic channels, recorded for the length of scan type 2
68 trace blocks total

Example 6.—The following is Example 6 again illustrating in graphical form the order of the trace blocks on tape. Note that even though auxiliary traces 1–4 are the same channels in both scan types, the data collected during the time of scan type 1 is recorded separately from those in scan type 2.

Scan type 1 is:

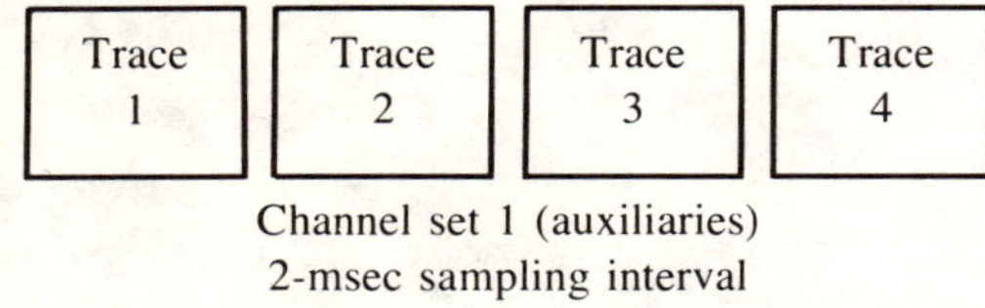

Channel set 1 (auxiliaries)
2-msec sampling interval

Trace | Trace 1 through 6 — Trace | Trace 1 through 6

Channel set 2
½ msec sampling
with low cut

Channel set 3
½ msec sampling
without low cut

Scan type 2 is:

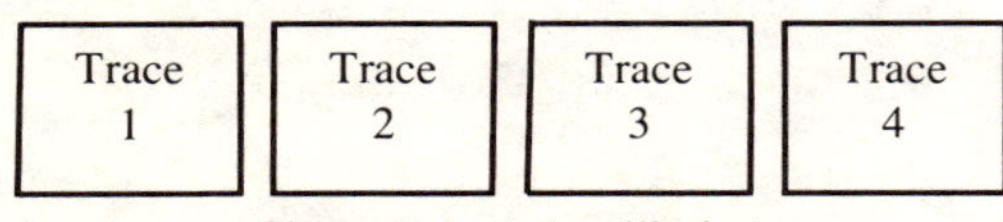

Channel set 1 (auxiliaries)
2-msec sampling

Trace 1 - - - - - - - through - - - - - - - Trace 48

Channel set 2
2-msec sampling

Note that there are no traces recorded for channel set 3 of scan type 2. There are 68 trace blocks total. Each new channel set starts with trace number 1.

REFERENCES

American National Standards Institute, 1973a, Recorded magnetic tape for information exchange, 800 BPI: ANSI X3.22-1973.

——— 1973b, Recorded magnetic tape for information exchange, 1600 BPI: ANSI X3.39-1973.

——— 1976, Recorded magnetic tape for information exchange, 6250 BPI: ANSI X3.54-1976.

Barry, K.M., Cavers, D.A., and Kneale, C.W., 1975, Recommended standards for digital tape formats: Geophysics, v. 40, p. 344-352.

Dampney, C.N.G., Funkhouser, D., and Alexander, M., 1978, Structure of the SEG point data exchange and field formats: Geophysics, v. 43, p. 216-227.

Meiners, E.P., Lenz, L.L., Dalby, A.E., and Hornsby, J.M., 1972, Recommended standards for digital tape formats: Geophysics, v. 37, p. 36-44.

Northwood, E.J., Weisinger, R.C., and Bradley, J.J., 1967, Recommended standards for digital tape formats: Geophysics, v. 32, p. 1073-1084.

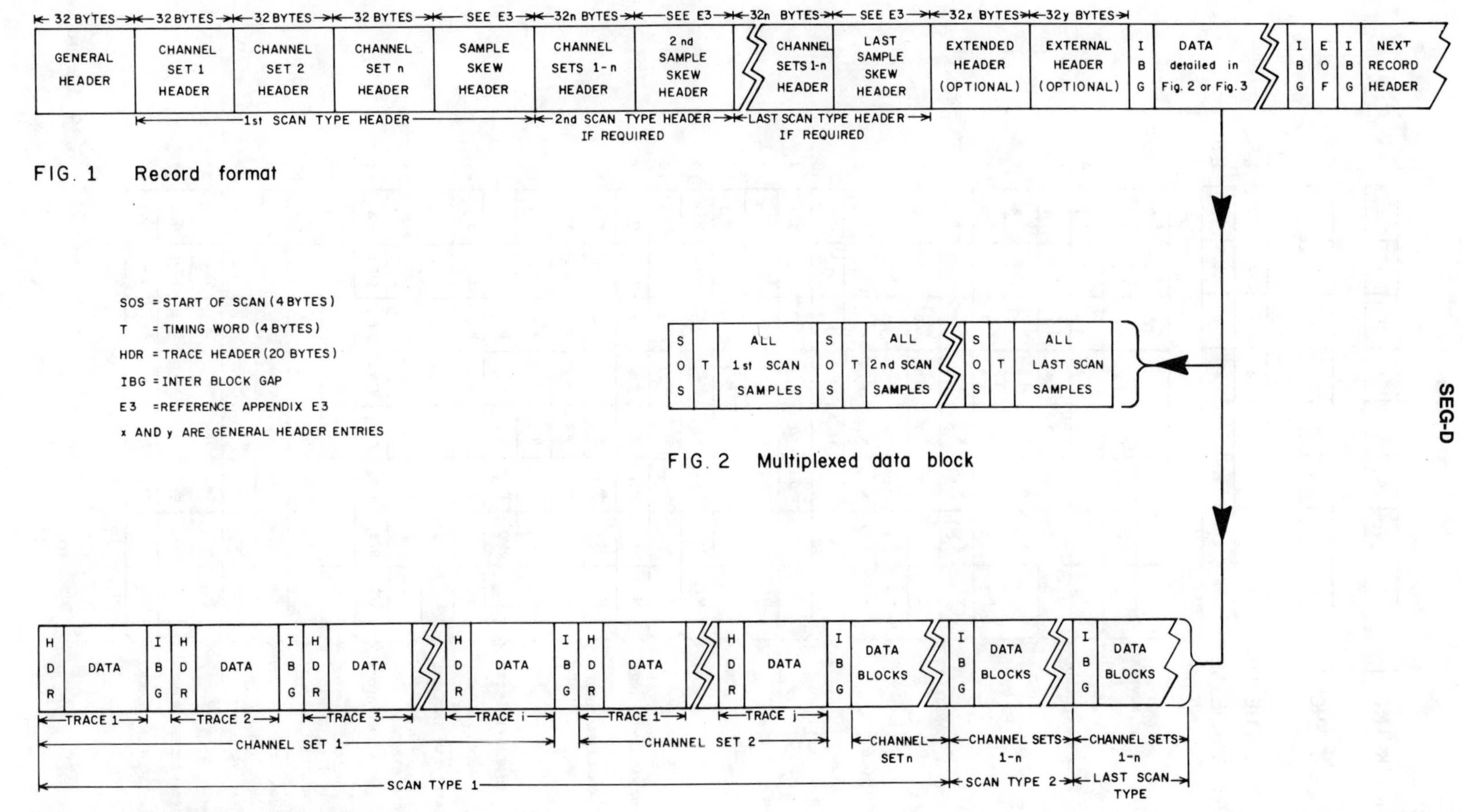

FIG. 1 Record format

FIG. 2 Multiplexed data block

FIG. 3 Demultiplexed data blocks

TRACK NO.	4	7	6	5	3	9	1	8	2	
BIT NO.	P	0	1	2	3	4	5	6	7	
BCD VALUE MSD		8	4	2	1	8	4	2	1	LSD
BINARY VALUE MSB		128	64	32	16	8	4	2	1	LSB

FILE NUMBER	F_1	F_1	F_1	F_1	F_2	F_2	F_2	F_2	1
	F_3	F_3	F_3	F_3	F_4	F_4	F_4	F_4	2
FORMAT CODE	Y_1	Y_1	Y_1	Y_1	Y_2	Y_2	Y_2	Y_2	3
	Y_3	Y_3	Y_3	Y_3	Y_4	Y_4	Y_4	Y_4	4
GENERAL CONSTANTS	K_1	K_1	K_1	K_1	K_2	K_2	K_2	K_2	5
	K_3	K_3	K_3	K_3	K_4	K_4	K_4	K_4	6
	K_5	K_5	K_5	K_5	K_6	K_6	K_6	K_6	7
	K_7	K_7	K_7	K_7	K_8	K_8	K_8	K_8	8
	K_9	K_9	K_9	K_9	K_{10}	K_{10}	K_{10}	K_{10}	9
	K_{11}	K_{11}	K_{11}	K_{11}	K_{12}	K_{12}	K_{12}	K_{12}	10
YEAR	YR_1	YR_1	YR_1	YR_1	YR_2	YR_2	YR_2	YR_2	11
DAY (DY)	0	0	0	0	DY_1	DY_1	DY_1	DY_1	12
	DY_2	DY_2	DY_2	DY_2	DY_3	DY_3	DY_3	DY_3	13
HOUR	H_1	H_1	H_1	H_1	H_2	H_2	H_2	H_2	14
MINUTE	MI_1	MI_1	MI_1	MI_1	MI_2	MI_2	MI_2	MI_2	15
SECOND	SE_1	SE_1	SE_1	SE_1	SE_2	SE_2	SE_2	SE_2	16
MANUFACTURER'S CODE	M_1	M_1	M_1	M_1	M_2	M_2	M_2	M_2	17
MANUFACTURER'S SERIAL NUMBER	M_3	M_3	M_3	M_3	M_4	M_4	M_4	M_4	18
	M_5	M_5	M_5	M_5	M_6	M_6	M_6	M_6	19
BYTES PER SCAN	B_1	B_1	B_1	B_1	B_2	B_2	B_2	B_2	20
	B_3	B_3	B_3	B_3	B_4	B_4	B_4	B_4	21
	B_5	B_5	B_5	B_5	B_6	B_6	B_6	B_6	22
BASE SCAN INTERVAL	I_3	I_2	I_1	I_0	I_{-1}	I_{-2}	I_{-3}	I_{-4}	23
POLARITY (P) SCANS/BLOCK EXPONENT (S/Bx)	P	P	P	P	S/B_{x3}	S/B_{x2}	S/B_{x1}	S/B_{x0}	24
SCANS/BLOCK	S/B_7	S/B_6	S/B_5	S/B_4	S/B_3	S/B_2	S/B_1	S/B_0	25
RECORD TYPE (Z)	Z	Z	Z	Z	R_1	R_1	R_1	R_1	26
RECORD LENGTH (R)	R_2	R_2	R_2	R_2	R_3	R_3	R_3	R_3	27
SCAN TYPES/RECORD	ST/R_1	ST/R_1	ST/R_1	ST/R_1	ST/R_2	ST/R_2	ST/R_2	ST/R_2	28
CHANNEL SETS /SCAN TYPE	CS_1	CS_1	CS_1	CS_1	CS_2	CS_2	CS_2	CS_2	29
SKEW BLOCKS	SK_1	SK_1	SK_1	SK_1	SK_2	SK_2	SK_2	SK_2	30
EXTENDED HEADER BLOCKS	EC_1	EC_1	EC_1	EC_1	EC_2	EC_2	EC_2	EC_2	31
EXTERNAL HEADER BLOCKS	EX_1	EX_1	EX_1	EX_1	EX_2	EX_2	EX_2	EX_2	32

FIG. 4. General header

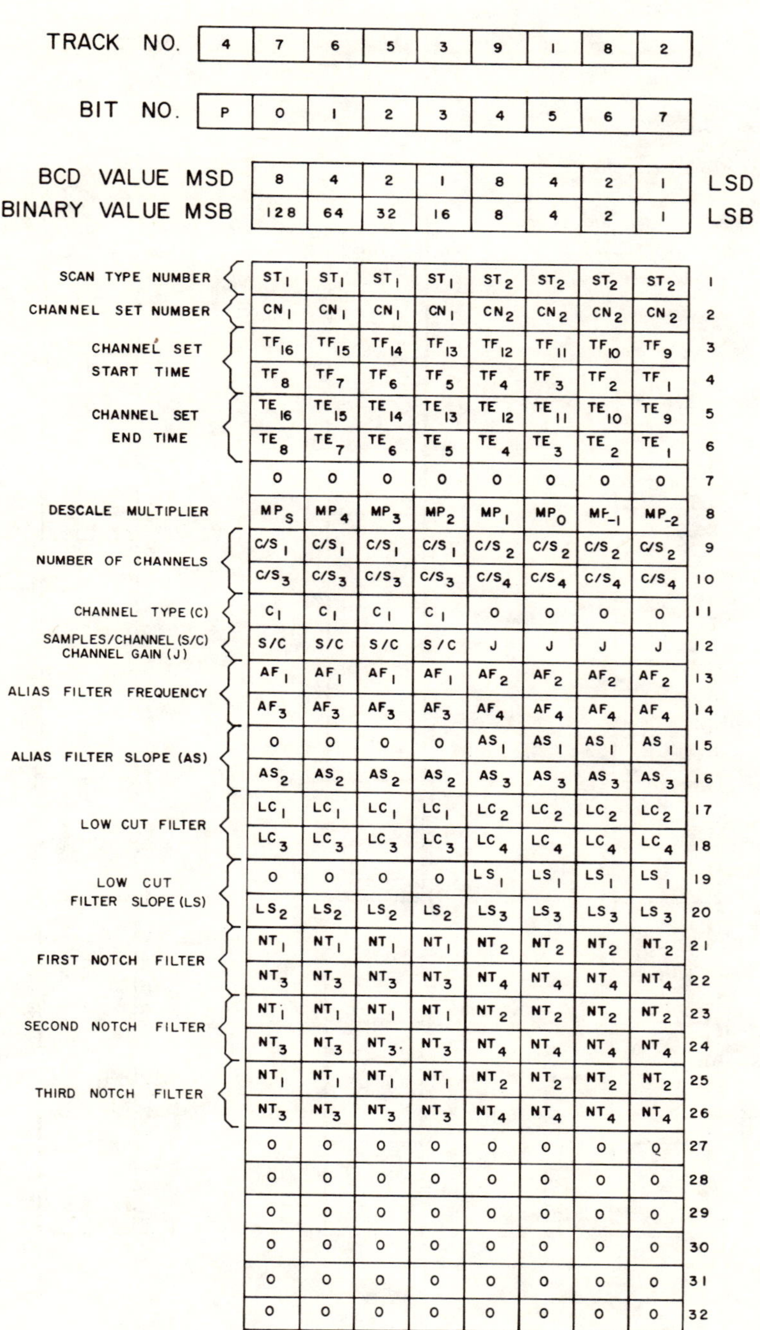

TRACK NO.	4	7	6	5	3	9	1	8	2	
BIT NO.	P	0	1	2	3	4	5	6	7	
BCD VALUE MSD		8	4	2	1	8	4	2	1	LSD
BINARY VALUE MSB		128	64	32	16	8	4	2	1	LSB

Field	0	1	2	3	4	5	6	7	Byte
SCAN TYPE NUMBER	ST_1	ST_1	ST_1	ST_1	ST_2	ST_2	ST_2	ST_2	1
CHANNEL SET NUMBER	CN_1	CN_1	CN_1	CN_1	CN_2	CN_2	CN_2	CN_2	2
CHANNEL SET START TIME	TF_{16}	TF_{15}	TF_{14}	TF_{13}	TF_{12}	TF_{11}	TF_{10}	TF_9	3
	TF_8	TF_7	TF_6	TF_5	TF_4	TF_3	TF_2	TF_1	4
CHANNEL SET END TIME	TE_{16}	TE_{15}	TE_{14}	TE_{13}	TE_{12}	TE_{11}	TE_{10}	TE_9	5
	TE_8	TE_7	TE_6	TE_5	TE_4	TE_3	TE_2	TE_1	6
	0	0	0	0	0	0	0	0	7
DESCALE MULTIPLIER	MP_S	MP_4	MP_3	MP_2	MP_1	MP_0	MP_{-1}	MP_{-2}	8
NUMBER OF CHANNELS	C/S_1	C/S_1	C/S_1	C/S_1	C/S_2	C/S_2	C/S_2	C/S_2	9
	C/S_3	C/S_3	C/S_3	C/S_3	C/S_4	C/S_4	C/S_4	C/S_4	10
CHANNEL TYPE (C)	C_1	C_1	C_1	C_1	0	0	0	0	11
SAMPLES/CHANNEL (S/C) CHANNEL GAIN (J)	S/C	S/C	S/C	S/C	J	J	J	J	12
ALIAS FILTER FREQUENCY	AF_1	AF_1	AF_1	AF_1	AF_2	AF_2	AF_2	AF_2	13
	AF_3	AF_3	AF_3	AF_3	AF_4	AF_4	AF_4	AF_4	14
ALIAS FILTER SLOPE (AS)	0	0	0	0	AS_1	AS_1	AS_1	AS_1	15
	AS_2	AS_2	AS_2	AS_2	AS_3	AS_3	AS_3	AS_3	16
LOW CUT FILTER	LC_1	LC_1	LC_1	LC_1	LC_2	LC_2	LC_2	LC_2	17
	LC_3	LC_3	LC_3	LC_3	LC_4	LC_4	LC_4	LC_4	18
LOW CUT FILTER SLOPE (LS)	0	0	0	0	LS_1	LS_1	LS_1	LS_1	19
	LS_2	LS_2	LS_2	LS_2	LS_3	LS_3	LS_3	LS_3	20
FIRST NOTCH FILTER	NT_1	NT_1	NT_1	NT_1	NT_2	NT_2	NT_2	NT_2	21
	NT_3	NT_3	NT_3	NT_3	NT_4	NT_4	NT_4	NT_4	22
SECOND NOTCH FILTER	NT_1	NT_1	NT_1	NT_1	NT_2	NT_2	NT_2	NT_2	23
	NT_3	NT_3	NT_3	NT_3	NT_4	NT_4	NT_4	NT_4	24
THIRD NOTCH FILTER	NT_1	NT_1	NT_1	NT_1	NT_2	NT_2	NT_2	NT_2	25
	NT_3	NT_3	NT_3	NT_3	NT_4	NT_4	NT_4	NT_4	26
	0	0	0	0	0	0	0	0	27
	0	0	0	0	0	0	0	0	28
	0	0	0	0	0	0	0	0	29
	0	0	0	0	0	0	0	0	30
	0	0	0	0	0	0	0	0	31
	0	0	0	0	0	0	0	0	32

FIG. 5 Channel set descriptor

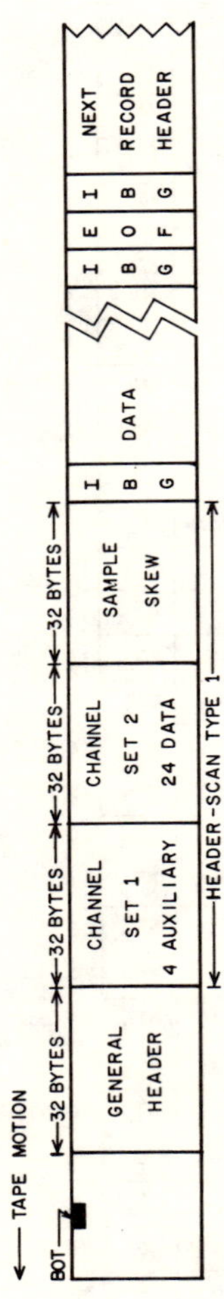

FIG. 6A Example 1

FIG. 6B Example 2

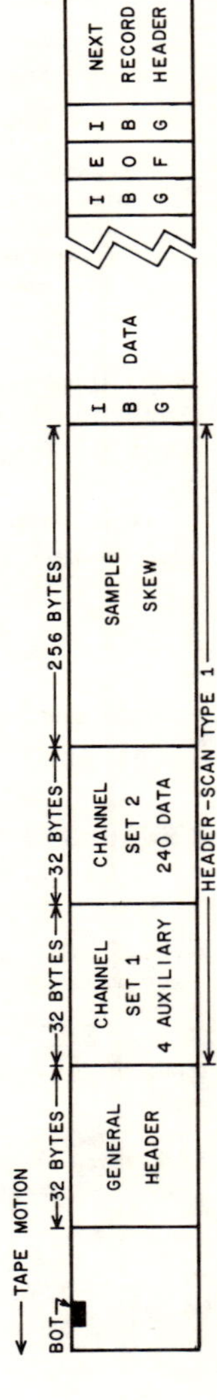

FIG. 6C Example 3

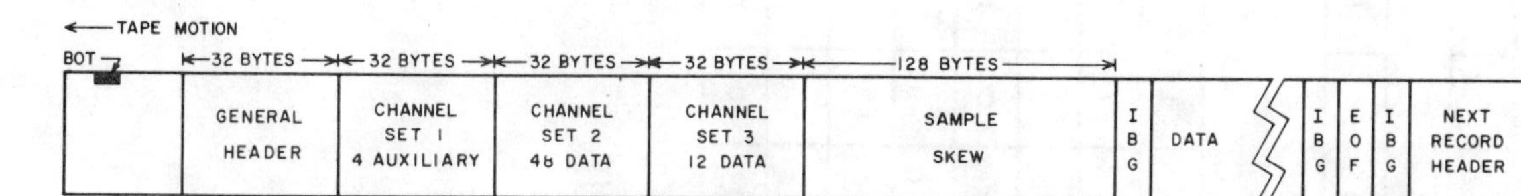

FIG. 7A Example 4

← TAPE MOTION
BOT
32 BYTES | 32 BYTES | 32 BYTES | 64 BYTES | 32 BYTES | 32 BYTES | 64 BYTES
GENERAL HEADER | CHANNEL SET 1 4 AUXILIARY | CHANNEL SET 2 12 DATA | SAMPLE SKEW | CHANNEL SET 1 4 AUXILIARY | CHANNEL SET 2 48 DATA | SAMPLE SKEW | IBG | DATA | IBG | EOF | IBG | NEXT RECORD HEADER
HEADER-SCAN TYPE 1
HEADER-SCAN TYPE 2

FIG. 7B Example 5

← TAPE MOTION
BOT
32 BYTES | 32 BYTES | 32 BYTES | 32 BYTES | 64 BYTES | 32 BYTES | 32 BYTES | 32 BYTES | 64 BYTES
GENERAL HEADER | CHANNEL SET 1 4 AUXILIARY | CHANNEL SET 2 6 DATA | CHANNEL SET 3 6 DATA | SAMPLE SKEW | CHANNEL SET 1 4 AUXILIARY | CHANNEL SET 2 48 DATA | CHANNEL SET 3 DUMMY | SAMPLE SKEW | IBG | DATA | IBG | EOF | IBG | NEXT RECORD HEADER
HEADER-SCAN TYPE 1
HEADER-SCAN TYPE 2

FIG. 7C Example 6

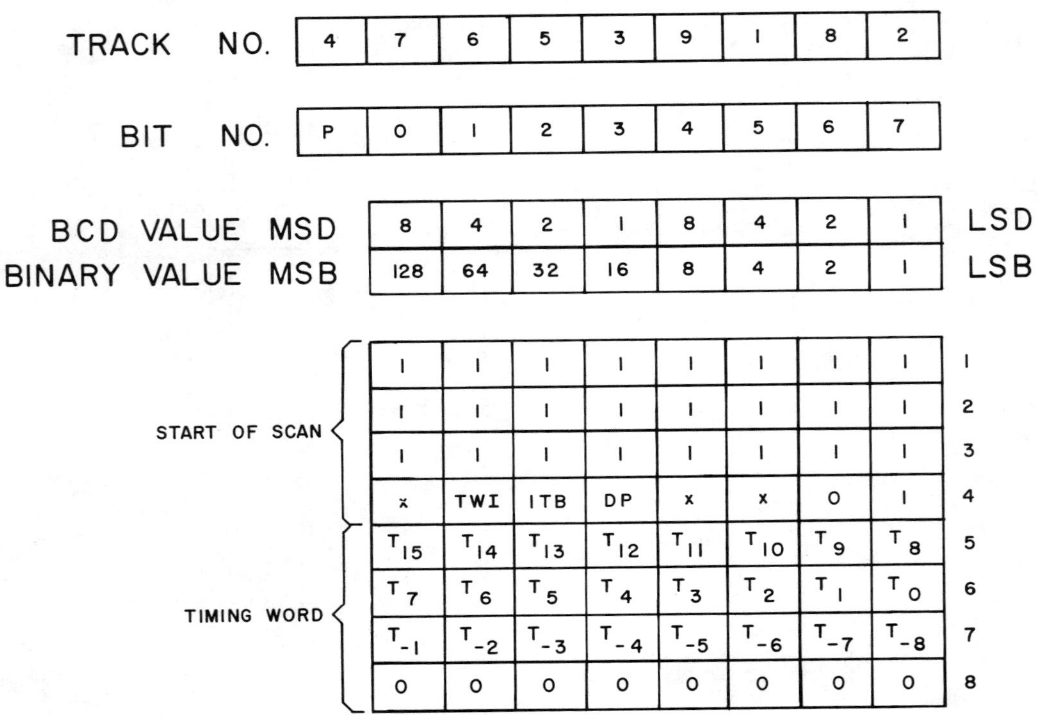

FIG. 8. Start of scan and timing word

Field									Byte
FILE NUMBER	F_1	F_1	F_1	F_1	F_2	F_2	F_2	F_2	1
	F_3	F_3	F_3	F_3	F_4	F_4	F_4	F_4	2
SCAN TYPE NUMBER	ST_1	ST_1	ST_1	ST_1	ST_2	ST_2	ST_2	ST_2	3
CHANNEL SET NUMBER	CN_1	CN_1	CN_1	CN_1	CN_2	CN_2	CN_2	CN_2	4
TRACE NUMBER	TN_1	TN_1	TN_1	TN_1	TN_2	TN_2	TN_2	TN_2	5
	TN_3	TN_3	TN_3	TN_3	TN_4	TN_4	TN_4	TN_4	6
FIRST TIMING WORD	T_{15}	T_{14}	T_{13}	T_{12}	T_{11}	T_{10}	T_9	T_8	7
	T_7	T_6	T_5	T_4	T_3	T_2	T_1	T_0	8
	T_{-1}	T_{-2}	T_{-3}	T_{-4}	T_{-5}	T_{-6}	T_{-7}	T_{-8}	9
	0	0	0	0	0	0	0	0	10
SAMPLE SKEW	SS_{-1}	SS_{-2}	SS_{-3}	SS_{-4}	SS_{-5}	SS_{-6}	SS_{-7}	SS_{-8}	11
	0	0	0	0	0	0	0	0	12
TIME BREAK WINDOW	TW_{15}	TW_{14}	TW_{13}	TW_{12}	TW_{11}	TW_{10}	TW_9	TW_8	13
	TW_7	TW_6	TW_5	TW_4	TW_3	TW_2	TW_1	TW_0	14
	TW_{-1}	TW_{-2}	TW_{-3}	TW_{-4}	TW_{-5}	TW_{-6}	TW_{-7}	TW_{-8}	15
	0	0	0	0	0	0	0	0	16
	0	0	0	0	0	0	0	0	17
	0	0	0	0	0	0	0	0	18
	0	0	0	0	0	0	0	0	19
	0	0	0	0	0	0	0	0	20

FIG.9 Demultiplexed trace header

APPENDIX A
HEADER BLOCK PARAMETERS

General header

All values are in packed BCD unless otherwise specified.

INDEX BYTE	ABBREVIATION	DESCRIPTION
1	F_1, F_2	File number of four
2	F_3, F_4	digits (0–9999)
3	Y_1, Y_2	Format code:
4	Y_3, Y_4	0015 20 bit binary multiplexed 0022 8 bit quaternary multiplexed 0024 16 bit quaternary multiplexed 0042 8 bit hexadecimal multiplexed 0044 16 bit hexadecimal multiplexed 0048 32 bit hexadecimal multiplexed 8015 20 bit binary demultiplexed 8022 8 bit quaternary demultiplexed 8024 16 bit quaternary demultiplexed 8042 8 bit hexadecimal demultiplexed 8044 16 bit hexadecimal demultiplexed 8048 32 bit hexadecimal demultiplexed 0200 Illegal, do not use 0000 Illegal, do not use
5	K_1, K_2	General constants, 12 digits
6	K_3, K_4	
7	K_5, K_6	
8	K_7, K_8	
9	K_9, K_{10}	
10	K_{11}, K_{12}	
11	YR_1, YR_2	Last two digits of year (0–99)
12	0, DY_1	Julian day 3 digits (1–366)
13	DY_2, DY_3	
14	H_1, H_2	Hour of day 2 digits (0–23) (Greenwich Mean Time)
15	MI_1, MI_2	Minute of hour 2 digits (0–59)
16	SE_1, SE_2	Second of minute 2 digits (0–59)
17	M_1, M_2	Manufacturer's code 2 digits Note: See Appendix B for the current assignments
18	M_3, M_4	Manufacturer's serial number, 4 digits
19	M_5, M_6	
20	B_1, B_2	Bytes per scan 6 digits (1–999,999) are utilized in the multiplexed formats to identify the number of bytes (including data, auxiliary, sync, and timing bytes, etc.) required to make up a complete scan. In a demultiplexed record, this field is not used and is recorded as zeros. (See Appendix E2)
21	B_3, B_4	
22	B_5, B_6	
23	I_3 thru I_{-4}	**Base scan interval.**—This is coded as a binary number with the LSB equal to 1/16 msec. This will allow sampling intervals from 1/16 through 8 msec

APPENDIX A—GENERAL HEADER

INDEX BYTE	ABBREVIATION	DESCRIPTION
		in binary steps. Thus, the allowable base scan intervals are 1/16, 1/8, 1/4, 1/2, 1, 2, 4, and 8 msec. The base scan interval is always the difference between successive timing words. Each channel used will be sampled one or more times per base scan interval.
24	P,	**Polarity.**—These 4 binary bits are measured on the sensors, cables, instrument, and source combination and are set into the system manually. The codes are: 0000 Untested 0001 Zero 0010 45 degrees 0011 90 degrees 0100 135 degrees 0101 180 degrees 0110 225 degrees 0111 270 degrees 1000 315 degrees 1001 1010 1011 1100 unassigned 1101 1110 1111[c]
	, S/BX_3 thru S/BX_0	This binary number (range 0 to 15) is an exponent of 2 and is used in conjunction with S/B (Byte 25).
25	S/B_7 thru S/B_0	This binary number (range 0 to 255) is used in conjunction with S/BX (see Byte 24) to indicate the number of scans in a block. If it is 0, the data body is one continuous block. Otherwise, the data body is composed of multiple blocks, each block containing $S/B \times 2^{S/BX}$ scans. It is valid only for multiplexed data.
26	Z,	Record type

Bits 0	1	2	3	
0	0	1	0	Test record
0	1	0	0	Parallel channel test
0	1	1	0	Direct channel test
1	0	0	0	Normal record
0	0	0	1	Other

INDEX BYTE	ABBREVIATION	DESCRIPTION
	, R_1	Record length from time zero (in increments of 0.5 times 1.024 sec). This value can be set from 00.5 to 99.5 representing times from 0.512 sec. to 101.888 sec. A setting of 00.0 indicates the record length is indeterminate.
27	R_2, R_3	

[c]Details of polarity codes and test methods are listed in the following reference: Thigpen, B. B., Dalby, A. E., Landrum, R., 1975, Special report of the subcommittee on polarity standards: Geophysics, v. 40, p. 694.

APPENDIX A—GENERAL HEADER

INDEX BYTE	ABBREVIATION	DESCRIPTION
28	ST/R_1, ST/R_2	Scan types per record. This 2 digit code is the number of scan types per record (1–99). (Zero is invalid.)
29	CS_1, CS_2	Number of channel sets per scan type (1–99). (Zero is invalid.) This 2 digit code is the number of channel sets per scan. If multiple scan types are used (such as in a switching sampling interval environment), this number is equal to the number of channel sets contained in the scan type with the largest number of channel sets. If scan types also exist with less than this maximum number of channel sets per scan type, dummy channel set descriptors will have to be recorded in the scan type header. This can be done by setting the number of channels in the dummy channel set descriptor to zero (reference Bytes 9 and 10 of the scan type header description). Example 6 illustrates this requirement.
30	SK_1, SK_2	Number of 32 byte fields added to the end of each scan type header in order to record the sample skew of all channels (0–99). (See Appendix E3). Zero indicates that skew is not recorded.
31	EC_1, EC_2	Extended header length. The extended header is used to record additional equipment parameters. An example of this would be parameters generated by the addition of a field stacker to the system. The two digits (0–99) in this field specify the number of 32 byte extensions.
32	EX_1, EX_2	External header length. The external header is used to record additional user supplied information in the header. The two digits (0–99) in this field specify the number of 32 byte extensions.

Scan type header (channel set descriptor)

The scan type header is determined by the system configuration and consists of one or more channel set descriptors each of 32 bytes followed by a series of 32 byte sample skew fields. A channel set is defined as a group of channels operating with the same set of parameters and being sampled as part of a scan of data. A scan type header can be composed of from 1 to 99 channel set descriptors. If dynamic parameter changes are required during the recording, additional scan type headers must be added, each containing the channel set descriptors necessary to define the new parameters. Each scan type header must have the same number of channel set descriptors (see Appendix E4 for header length calculation).

APPENDIX A—CHANNEL SET DESCRIPTOR

INDEX BYTE	ABBREVIATION	DESCRIPTION
1	ST_1, ST_2	These two digits (1–99) identify the number of the scan type header to be described by the subsequent bytes. The first scan type header is 1 and the last scan type header number is the same value as Byte 28 (ST/R) of the general header. If a scan type header contains more than one channel set descriptor, the scan type header number will be repeated in each of its channel set descriptors. If the system does not have dynamic parameter changes during the record, such as switched sampling intervals, there will only be one scan type header required.
2	CN_1, CN_2	These two digits (1–99) identify the channel set to be described in the next 30 bytes within this scan type header. The first channel set is "1" and the last channel set number is the same number as Byte 29 (CS) of the general header. If the scan actually contains fewer channel sets than CS, then dummy channel set descriptors are included as specified in Byte 29 of general header.
3 4	TF_{16} thru TF_9 TF_8 thru TF_1	Channel set starting time. This is a binary number where $TF_1 = 2^1$ msec (2-msec increments). This number identifies the timing word of the first scan of data in this channel set. In a single scan type record, this would typically be recorded as a zero (an exception might be deep water recording). In multiple scan type records, this number represents the starting time, in milliseconds, of the channel set. Start times from 0 to 131,070 msec (in 2-msec increments) can be recorded.
5 6	TE_{16} thru TE_9 TE_8 thru TE_1	Channel set end time. This is a binary number where $TE_1 = 2^1$ milliseconds (2 millisecond increments). These two bytes represent the record end time of the channel set in milliseconds. In a multiplexed record, all channels of a channel set must be of the same length. TE may be used in a demultiplexed record to allow the termination of a particular channel set shorter than other channel sets within its scan type. In a single scan type record, Bytes 5 and 6 would be the length of the record. End times up to 131,070 msec (in 2-msec increments) can be recorded.
7	0, 0	
8	MP_s, MP_4 thru MP_{-2}	This sign magnitude binary number is the exponent of the base 2 multiplier to be used to descale the data on tape to obtain input voltage in millivolts. The radix point is between MP_0 and MP_{-1}. This multiplier has a range of $2^{31.75}$ to $2^{-31.75}$. (See Appendix E7.)
9 10	C/S_1, C/S_2 C/S_3, C/S_4	This is the number of channels in this channel set. It can assume a number from 0–9999.

APPENDIX A—CHANNEL SET DESCRIPTOR

INDEX BYTE	ABBREVIATION	DESCRIPTION
11	C_1, 0	Channel type identification:

Bit 0	1	2	3	
0	1	1	1	Other
0	1	1	0	External data
0	1	0	1	Time counter[d]
0	1	0	0	Water break
0	0	1	1	Up hole
0	0	1	0	Time break
0	0	0	1	Seis
0	0	0	0	Unused
1	0	0	0	Signature, unfiltered
1	0	0	1	Signature, filtered

INDEX BYTE	ABBREVIATION	DESCRIPTION
12	S/C,	This packed BCD number is an exponent of 2. The number ($2^{S/C}$) represents the number of subscans of this channel set in the base scan. Possible values for this parameter ($2^{S/C}$) are 1 to 512 (2^0 to 2^9). Reference Byte 23 of the general header.)
12	, J	Channel gain control method.

Bits 4	5	6	7	Gain mode
0	0	0	1	- (1) Individual AGC
0	0	1	0	- (2) Ganged AGC
0	0	1	1	- (3) Fixed gain
0	1	0	0	- (4) Programmed gain
1	0	0	0	- (8) Binary gain control
1	0	0	1	- (9) IFP gain control

INDEX BYTE	ABBREVIATION	DESCRIPTION
13 14	AF_1, AF_2 AF_3, AF_4	Alias filter frequency. It can be coded for any frequency from 0 to 9999 Hz.
15 16	0, AS_1 AS_2, AS_3	Alias filter slope in dB per octave. It can be coded from 0 to 999 dB in 1-dB steps. A zero indicates the filter is out (see Appendix E5 for definition).
17 18	LC_1, LC_2 LC_3, LC_4	Low-cut filter setting. It can be coded for any frequency from 0 to 9999 Hz.
19 20	0, LS_1 LS_2, LS_3	Low-cut filter slope. It can be coded for any slope from 0 to 999 dB per octave. A zero slope indicates the filter is out. (See Appendix E5 for definition.)
21 22	NT_1, NT_2 NT_3, NT_4	Notch frequency setting. It can be coded for any frequency from 0 to 999.9 Hz. The out filter is written as 000.0 Hz.

[d]Illegal code for this format because the timing counter is part of the start of scan and cannot be identified as part of a channel.

APPENDIX A—CHANNEL SET DESCRIPTOR

INDEX BYTE	ABBREVIATION	DESCRIPTION

The following notch filters are coded in a similar manner:

INDEX BYTE	ABBREVIATION	DESCRIPTION
23	NT_1, NT_2	Second notch frequency
24	NT_3, NT_4	
25	NT_1, NT_2	Third notch frequency
26	NT_3, NT_4	
27		
28		
29	Unused. Written as zeros.	
30		
31		
32		

APPENDIX B
MANUFACTURERS OF SEISMIC DIGITAL FIELD RECORDERS

Code No.

01 Alpine Geophysical Associates, Inc. (Obsolete)[e]
65 Oak Street
Norwood, New Jersey

02 Applied Magnetics Corporation (See 09)
75 Robin Hill Rd.
Goleta, California 93017

05 Dyna-Tronics Mfg. Corporation (Obsolete)[e]
5820 Star Lane
Box 22202
Houston, Texas 77027

06 Electronic Instrumentation, Inc. (Obsolete)[e]
601 Dooley Road
Box 34046
Dallas, Texas 75234

07 Electro-Technical Labs
Div. of Geosource, Inc.
6909 Southwest Freeway
Box 36827
Houston, Texas 77036

08 Fortune Electronics, Inc. (Obsolete)[e]
5606 Parkersburg Drive
Houston, Texas 77036

09 Geo Space Corporation (Subsidiary of Applied Magnetics Company)
5803 Glenmont Drive
Box 36374
Houston, Texas 77036

17 GUS Manufacturing, Inc.
P.O. Box 10013
El Paso, Texas 79991

18 Input/Output, Inc.
8009 Harwin Dr.
Houston, Texas 77036

Code No.

10 Leach Corporation (Obsolete)[e]
405 Huntington Drive
San Marino, California

03 Litton Resources Systems, Inc.
3930 Westholme Dr.
Houston, Texas 77063

11 Metrix Instrument Co. (Obsolete)[e]
8200 Westglen
Box 36501
Houston, Texas 77063

12 Redcor Corporation (Obsolete)[e]
7800 Deering Avenue
Box 1031
Canoga Park, California 91304

14 Scientific Data Systems (SDS) (Obsolete)[e]
1649 Seventeenth Street
Santa Monica, California 90404

13 Sercel (Societe d'Etudes, Recherches Et Constructions Electroniques)
25 X, 44040 Nantes Cedex,
France

04 SIE, Inc.
5110 Ashbrook
Box 36293
Houston, Texas 77036

15 Texas Instruments, Inc.
P.O. Box 1444
Houston, Texas 77001

[e]This was originally extracted from Geophysics, v. 32. The ones marked obsolete do not appear in the latest edition of the Geophysical Directory (1979). Should additional manufacturer code numbers be required, contact the SEG Standards Committee for the assignment of these numbers.

APPENDIX C—GLOSSARY

Base scan interval The time between timing words. A base scan interval usually contains one scan but under some conditions may contain multiple subscans.

Block The data between gaps on tape.

Channel set One or more channels sampled at the same sampling interval and containing the same filter, fixed gain, and other fixed parameter information.

Channel set descriptor A unit of the scan type header describing the parameters of a channel set.

Data recording method The arrangement of bits to represent samples on tape.

File All data recorded from a single energy impulse or sweep. It may also be the sum of a number of energy impulses or sweeps. Literally, it is all of the blocks between file marks.

Format Data recording method combined with a multiplexed/demultiplexed indicator (see general header Bytes 3 and 4).

General header The first header in the header block. It contains information common to the entire record.

Index byte The byte number of some particular parameter within the general or scan type header.

Packed BCD Binary coded decimal digits represented by four data bits.

Sample skew The fraction of the base scan interval between the timing word and the actual time the sample was taken in a base scan interval (not related to position on tape).

Sampling interval The interval between readings such as the time between successive samples of a digital seismic trace.

Scan One complete sequence of events, such as sampling all channels. Data recorded during a base scan interval.

Scan interval The interval between readings of all samples contained in a scan type.

Scan type One complete set of channel sets which make up a scan. A seismic record contains multiple scans, and may or may not contain more than one scan type.

Scan type header A header containing one or more channel set descriptors and the skew information.

Subscan A set of samples containing one sample for each channel in a channel set.

Time break window Time interval in which time break is expected. If time break does not occur by the end of the window, internal time break is generated.

Trace A record of one seismic channel within a scan type. A collection of a sequential set of points from one seismic channel.

Trace block A block containing the data of one trace or a part of a trace with constant parameters.

APPENDIX D—HEADER DESCRIPTORS

G = general header S = scan type header (channel set descriptor)

ABBREVIATION	HEADER	BYTE NO.	DESCRIPTION
AF	S	13, 14	ALIAS FILTER FREQUENCY
AS	S	15, 16	ALIAS FILTER SLOPE
B	G	20, 21, 22	BYTES PER SCAN (MULTIPLEXED ONLY)
BCD	—		BINARY CODED DECIMAL
BOT			BEGINNING OF TAPE MARK
C	S	11	CHANNEL TYPE IDENTIFICATION
CN	S	2	CHANNEL SET NUMBER
CS	G	29	CHANNEL SETS PER SCAN TYPE
C/S	S	9, 10	CHANNELS IN THIS CHANNEL SET
DP	—		DYNAMIC PARAMETER CHANGE BIT (SEE THE MULTIPLEXED DATA BLOCK, SOS BYTE 4)
DY	G	12, 13	DAY OF YEAR
EC	G	31	EXTENDED HEADER LENGTH
EOF	—		END OF FILE MARK
EOT			END OF TAPE MARK
EX	G	32	EXTERNAL HEADER LENGTH
F	G	1, 2	FILE NUMBER
H	G	14	HOUR OF DAY
HDR	—		HEADER FOR DEMULTIPLEXED TRACE
HL	—		HEADER LENGTH (SEE APPENDIX E4)
I	G	23	BASE SCAN INTERVAL
IBG	—		INTERBLOCK GAP (ALSO GAP)
ITB	—		INTERNAL TIME BREAK (SEE THE MULTIPLEXED DATA BLOCK, SOS BYTE 4)
J	S	12	GAIN CONTROL METHOD
K	G	5 THRU 10	GENERAL CONSTANTS
LC	S	17, 18	LOW CUT FILTER FREQUENCY
LS	S	19, 20	LOW CUT FILTER SLOPE
LSB	—		LEAST SIGNIFICANT BIT
LSD	—		LEAST SIGNIFICANT DIGIT
M	G	17 THRU 19	MANUFACTURER'S CODE AND SERIAL NUMBER
MI	G	15	MINUTE OF HOUR
MP	S	8	DESCALING EXPONENT
MSB	—		MOST SIGNIFICANT BIT
MSD	—		MOST SIGNIFICANT DIGIT
NT	S	21 THRU 26	NOTCH FILTER FREQUENCY
P	G	24	POLARITY
R	G	26, 27	RECORD LENGTH
S	—		SIGN BIT
S/B, S/BX	G	24, 25	NUMBER OF SCANS PER BLOCK

APPENDIX D—HEADER DESCRIPTORS

ABBREVIATION	HEADER	BYTE NO.	DESCRIPTION
S/C	S	12	EXPONENT OF SAMPLES PER CHANNEL IN THE BASE SCAN
SE	G	16	SECOND OF MINUTE
SK	G	30	NUMBER OF 32 BYTE SKEW FIELDS
SOS	—		START OF SCAN (MULTIPLEXED DATA BLOCK)
SS	—		SAMPLE SKEW
S/S	—		SAMPLES/SCAN
ST	S	1	SCAN TYPE NUMBER
ST/R	G	28	SCAN TYPES PER RECORD
T	—		TIMING WORD (MULTIPLEXED DATA BLOCK)
TF	S	3, 4	FIRST TIMING WORD IN THIS CHANNEL SET
TE	S	5, 6	END TIME OF THIS CHANNEL SET
TN	—		DEMULTIPLEXED TRACE NO. (SEE TRACE HEADER)
TW	—		TIME BREAK WINDOW (SEE DEMULTIPLEXED DATA BLOCK, TRACE HEADER BYTES 13, 14 AND 15)
TWI	—		TIME BREAK WINDOW INDICATOR (SEE MULTIPLEXED DATA BLOCK, SOS BYTE 4)
Y	G	3, 4	FORMAT CODE (DATA RECORDING METHOD)
YR	G	11	YEAR (LAST TWO DIGITS)
Z	G	26	RECORD TYPE

APPENDIX E
SAMPLE CALCULATIONS

E1 Samples per scan type

$$S/S = \sum_{1}^{CS} C/S \times 2^{S/C}$$

where

S/S = samples per scan type
C/S = channels in this channel set (channel set descriptor Bytes 9 and 10)
$2^{S/C}$ = samples per channel (in this channel set) (channel set descriptor Byte 12)
CS = number of channel sets in this scan type (general header Byte 29)

For example, for a 2-msec base scan interval with 4 auxiliary channels at 2 msec, 96 channels at 2 msec and 12 channels at ½ msec. There are three channel sets, so CS = 3.

$$S/S = C/S \times 2^{S/C}\Big|_{CS=1} + C/S \times 2^{S/C}\Big|_{CS=2} + \ldots$$

$$S/S = 4 \times 1 + 96 \times 1 + 12 \times 4$$

$$S/S = 4 + 96 + 48 = 148$$

Note that all scan types must have the same number of data samples.

E2 Bytes per scan

$$B = 8 + \sum_{1}^{CS} C/S \times 2^{S/C} \times \text{bytes/sample}$$

where

B = bytes per scan
8 = start of scan plus timing word
C/S = channels in this channel set (channel set descriptor Bytes 9 and 10)
$2^{S/C}$ = samples per channel (in this channel set) (channel set descriptor header Byte 12)
CS = number of channel sets in this scan type (general header Byte 29)

Bytes/sample	format code (general header Bytes 3 and 4)
1.0	0042, 0022
2.0	0024, 0044
2.5	0015
4.0	0048

For example, for a 2-msec base scan with a 2.5 byte format that contains 4 auxiliary channels, 96 channels at 2 msec and 12 channels at ½ msec.

$$B = 8 + (4 \times 1 \times 2.5) + (96 \times 1 \times 2.5) + (12 \times 4 \times 2.5)$$

$$B = 8 + (4 + 96 + 48) \times 2.5$$

$$B = 8 + 148 \times 2.5 = 378 \text{ bytes}$$

E3 Skew fields per scan type

$$SK = \frac{S/S}{32}$$

(If the quotient is not a whole number, round up to the next largest whole number.)

where

SK = skew fields (of 32 bytes each) per scan type (general header Byte 30)
S/S = samples per scan (Appendix E1)

Substituting for S/S from Appendix E1:

$$SK = \frac{\sum_{1}^{CS} C/S \times 2^{S/C}}{32}$$

(If the quotient is not a whole number, round up to the next largest whole number.)

where

CS = the number of channel sets in each scan type (general header Byte 29)

C/S = channels in this channel set (channel set descriptor Bytes 9 and 10)

$2^{S/C}$ = samples per channel in this channel set (channel set descriptor Byte 12).

For example, for a 2-msec base scan with 4 auxiliary channels at 2 msec, 96 channels at 2 msec and 12 channels at ½ msec

$$SK = \frac{4 \times 1 + 96 \times 1 + 12 \times 4}{32}$$

$$= \frac{148}{32} = 4\,\frac{20}{32} \text{ roundup} = 5 \text{ fields of 32 bytes each}$$

E4 Total header length

$$HL = 32 \times [ST/R\,(CS + SK) + 1 + EC + EX],$$

where

HL = header length (bytes)

ST/R = number of scan types per record (general header Byte 28)

CS = number of channel sets per scan type (general header Byte 29)

SK = skew fields per scan type (general header Byte 30)

EC = extended header length (general header Byte 31)

EX = external header length (general header Byte 32)

Example 1: For a system with a 2-msec base scan, 4 auxiliary channels at 2 msec, 96 channels at 2 msec, and 12 channels at ½ msec

$HL = 32 \times (1 \times (3 + 5) + 1 + EC + EX)$

ST/R = 1 since there is only one scan type in this example,

$= 32 \times (9) = 288$ bytes + extended header + external header

E5 Filter slope calculation

Modern filters may not have a constant slope, so it is necessary to define this parameter. The slope is defined as the asymptote of effective performance as it would be in a constant slope filter. This slope is zero dB attenuation at the cut-off frequency and a specific attenuation at the beginning of the stop band. The chosen values are 40 dB for a low-cut filter and 60 dB for an anti-alias filter.

Low-cut filter slope calculation.—

$$LS = \frac{40}{\log_2 f_{LCO}/f_{40}} = \frac{40}{3.322 \log_{10} f_{LCO}/f_{40}} = \frac{12.04}{\log_{10} f_{LCO}/f_{40}},$$

LS = low-cut filter slope (channel set descriptor Bytes 19 and 20),

f_{40} = the frequency of 40 dB low-cut filter attenuation,

f_{LCO} = low-cut filter cut-off frequency usually 6 or 12 dB attenuation.

Alias-filter slope calculation.—

$$AS = \frac{60}{\log_2 f_{60}/f_{ACO}} = \frac{60}{3.322 \log_{10} f_{60}/f_{ACO}} = \frac{18.06}{\log_{10} f_{60}/f_{ACO}},$$

AS = alias filter slope (channel set descriptor Bytes 15 and 16)

f_{60} = the frequency of 60 dB alias-filter attenuation

f_{ACO} = alias-filter cut-off frequency usually 3 or 6 dB attenuation

The resultant slope in the above calculations is rounded to the nearest whole number and is written in the channel set descriptor.

E6 Calculation of byte offset from the beginning of the header block to a specific byte index in channel set descriptor *i* of scan type header *j*

$$\text{Byte no.} = 32 + \text{index byte} + 32\,(CN_i - 1) + 32\,(ST_j - 1)\,(CS + SK)$$

where

CN_i = channel set number of interest

index byte = byte number of interest in the channel set descriptor

CS = number of channel sets per scan type (general header Byte 29).

ST_j = scan type of interest

SK = skew fields per scan type (Appendix E3)

$1 \leq CN_i \leq CS$

$1 \leq ST_j \leq ST/R$; $i = 1, 2, \ldots, ST/R$

Example: For a 2 msec base scan system having 2 scan types: (1) 4 auxiliary channels at 2 msec plus 12 channels at ½ msec, and in the second scan type (2) 4 auxiliary channels at 2 msec plus 48 channels at 2 msec.

First calculate the samples per scan (S/S) per Appendix E1 and skew field (SK) per Appendix E3.

$$S/S = \sum_{1}^{CS} C/S \times 2^{S/C}$$

$$S/S = C/S \times 2^{S/C} \Big|_{CS=1} + C/S \times 2^{S/C} \Big|_{CS=2}$$

$$= 4 \times 1 + 12 \times 4$$

$$S/S = 4 + 48 = 52.$$

Note: It is not necessary to evaluate the samples per scan in the second scan type because all scan types must have the same number of samples.

Second, evaluate the skew fields per scan per Example 3.

$$SK = \frac{\sum_{1}^{CS} S/S}{32} \quad \text{(round up to the next largest whole number),}$$

$$SK = \frac{52}{32} = 1\frac{20}{32} \quad \text{rounded up to 2.}$$

Now, compute the header byte number of index Byte 11 of the second channel set in the second scan type.

Byte no. = 32 + 11 + 32 (2 − 1) + 32 (2 − 1) (2 + 2)

Byte no. = 43 + 32 (1) + 32 (4)

Byte no. = 203.

E7 Calculation and use of MP

The MP parameter is provided to allow the dimensionless numbers recorded on tape to be "descaled" back to the instantaneous sample values in millivolts at the system inputs. MP is encoded in Byte 8 of each channel set descriptor in the scan type header. It is a sign and magnitude binary exponent. It can have any value between −31.75 and +31.75 in increments of .25.

In general, recording systems scale the input signal level in order to match the useful range of input levels to the gain-ranging amplifier. MP must account for all scaling (unless, as in the 4 byte hexadecimal case, the data recording method has sufficient range).

MP CALCULATION

The calculation of MP for a data recording method is given by one of the following equations:

(1) MP = FS − PA − Cmax; for binary exponents,

(2) MP = FS − PA − 2xCmax; for quaternary exponents,

(3) MP = FS − PA − 4xCmax; for hexadecimal exponents (except the 4 byte excess 64 method),

(4) MP = FS − PA − 4 (Cmax − 64); for excess 64 hexadecimal exponents,

where

2^{FS} = Converter full scale (millivolts),

2^{PA} = Minimum system gain,

and

Cmax = maximum value of the data exponent.

Cmax = 15 for binary exponents,

= 7 for quaternary exponents,

= 3 for hexadecimal exponents except excess 64; and

= 127 for excess 64 exponents.

The term "minimum system gain" includes preamplifier gain and the minimum floating point amplifier gain. For example, one system may use a preamplifier gain of 256 and a minimum floating point amplifier gain of one. The minimum system gain is $256 \times 1 = 2^8$, so PA = 8. Another system may use a preamplifier gain of 320 and a minimum floating point amplifier gain of 0.8. In this case, the minimum system gain is $320 \times 0.8 = 256$ or 2^8. Again PA = 8.

PA may also account for any amplification needed to accommodate an analog to digital converter with a full scale value that is not a power of 2 in millivolts. For example, a 10 V (10,000 mV) converter may be preceded by an amplifier with a gain of 1.221 (10,000/8,192). This gain may be accounted for in PA. Alternatively, it could be considered part of the converter, making it appear to have a binary full scale. In either case, FS − PA must be a multiple of 0.25.

JUSTIFICATIONS FOR THE EQUATIONS

The output of the analog-to-digital converter is written as the fractional portion of the data value. This is equivalent to dividing the value by the full scale of the converter. In order to compensate for this, the data value recorded on tape must be multiplied by the full scale value of the converter (2^{FS}). Thus FS appears in equations (1)–(4) with a positive sign.

The input signal was multiplied by the minimum system gain (2^{PA}) which, as mentioned, includes any preamplification gain, minimum floating point amplifier gain, or analog-to-digital converter adjustment gain. The data recorded on tape must be divided by this minimum system gain; thus, PA appears in the equations with a negative sign.

Large input signals converted at minimum floating point amplifier gain are written on tape with the maximum exponent for the data recording method used. Likewise, small signals converted at full gain are written with the minimum exponent. The data as written have been multiplied by the exponent base raised to Cmax (or Cmax −64 in the excess 64 case). Thus Cmax appears in the equations with a negative sign. MP is a power of 2 so the quaternary and hexadecimal Cmax values are multiplied by 2 and 4, respectively ($4^C = 2^{2C}$ and $16^C = 2^{4C}$).

EXAMPLES

Note: In the following examples, all logarithms are base 10.

Example 1

Assume: quaternary data recording method, converter full scale = 8192 mV, preamplifier gain = 320, and minimum floating point amplifier gain = 0.8.

Then

Cmax = 7,

FS = log 8192/log 2 = 13,

PA = log (320 × 0.8)/log 2 = 8.

and

$$MP = 13 - 8 - 2 \times 7 = -9.$$

Example 2

Assume: binary data recording method, converter full scale = 10 volts, preamplifier gain = 128, and minimum floating point amplifier gain = 1.0.

Then

Cmax = 15,

FS = log 10000/log 2 = 13.287 . . . ,

PA = log 128/log 2 = 7,

and

$$MP = 13.287\ldots -7 - 15 = -8.712\ldots.$$

But MP must be a multiple of 0.25. This can be achieved by changing the preamplifier gain to 131.39. Then PA = log 131.39/log 2 = 7.037 . . . . , and MP = 13.287 . . . −7.037 . . . −15 = −8.75.

MP could be made a whole number by changing the preamplifier gain to 156.25. Then PA = log 156.25/ log 2 = 7.287 . . . . , and MP = 13.287 . . . −7.287 . . . −15 = −9.

Note that this is equivalent to preceding the converter with an amplifier of gain 1.220 . . . thus making the converter appear to have a full scale value of 8192 mV.

Then

FS = log 8192/log 2 = 13,

PA = log 128/log 2 = 7,

and

$$MP = 13 - 7 - 15 = -9.$$

Example 3

Assume: hexadecimal (1 or 2 byte) data recording method, converter full scale = 8192 mV, preamplifier gain = 256, and minimum floating point amplifier gain = 1.0.

Then

Cmax = 3,

FS = log 8192/log 2 = 13,

PA = log 256/log 2 = 8,

and

$$MP = 13 - 8 - 4 \times 3 = -7.$$

Note: In the 4 byte hexadecimal case, it is expected that MP will generally be recorded as zero because of the large range of the number system of this data recording method. Nonetheless, equation 4 must still be valid:

$$MP = FS - PA - 4x(Cmax - 64).$$

If MP = 0, then

$FS - PA = 4(Cmax - 64)$.

Since Cmax is always an integer, FS − PA must be a multiple of 4.

Note: The data recording method (binary, quaternary, or hexadecimal) does not indicate the type of floating point amplifier used. Any gain ranging amplifier and converter outputs can be converted to any of the data recording methods.

USE OF MP

As indicated in the descriptions of the data recording formats, MP is applied by multiplying the recorded data by 2^{MP}.

E8 Calculation of the byte offset from the beginning of the header block to the skew information byte for channel *k* of channel set *i* of scan type *j*

For clarity, the equation is given on several lines. The terms on each line are explained.

SS OFFSET = 32	(general header)
$+ 32(ST_j - 1)(CS + SK)$	(all previous scan type headers)
$+ 32(CS)$	(channel set descriptors in scan type header *j*)
$+ \sum_{1}^{CN_i - 1} C/S \times 2^{S/C}$	(samples in the base scan for all previous channel sets in this scan type)
$+ k$	(position of channel in this channel set)

where

ST_j = scan type of interest,
CS = number of channel sets per scan type (general header Byte 29),
SK = number of skew fields per scan type (Appendix E3),
CN_i = channel set number of interest,
C/S = channels in this channel set (channel set descriptor Bytes 9 and 10), and
$2^{S/C}$ = samples per channel of this channel set in a base scan (channel set descriptor Byte 12).

Note that if there are multiple subscans of a channel set within a base scan (i.e., $2^{S/C}$ greater than one), a skew information byte is written for each sample in each subscan.

The above equation accounts for multiple subscans of previous channel sets, but it determines the location of the skew information byte of channel *k* in the first subscan only of channel set *i*. Skew information for subsequent subscans can be obtained in two ways:

(1) Add multiples of C/S to the computed offset to obtain the offset to the byte of interest; or
(2) Add multiples of $2^{-S/C}$ to the value of the skew information byte obtained for the first subscan.

For example, assume there are two scan types with three channel sets, as follows:

Scan type 1:
- channel set 1—4 channels at 4 msec (1 subscan per base scan)
- channel set 2—24 channels at 2 msec (2 subscans per base scan)
- channel set 3—12 channels at 1msec (4 subscans per base scan)

Scan type 2:
- Channel set 1—4 channels at 4 msec (1 subscan per base scan)
- Channel set 2—48 channels at 2 msec (2 subscans per base scan)
- Channel set 3—0 channels (dummy).

The offset to the skew information byte for channel 11 of channel set 2 in scan type 2 is computed below. The number of skew fields (Appendix E3) is:

$$\frac{\sum_{1}^{CS} C/S \times 2^{S/C}}{32} = 4 + 24 \times 2 + 12 \times 4$$

$$= \frac{100}{32}$$

or 4 (rounded up).

$$\text{SS offset} = 32 + 32(ST_j - 1)(CS + SK) + 32(CS)$$

$$+ \sum_{1}^{CN_i - 1} C/S \times 2^{S/C} + k$$

$$= 32 + 32(2 - 1)(3 + 4) + 32(3) + 4 + 11$$

$$= 367.$$

The position of the skew information byte for the second subscan of channel 11 in this scan is:

$$367 + 48 = 415.$$

E9 Calculation of the number of bytes per block for a gapped multiplexed record

$$\text{Bytes per block} = (S/B \times 2^{S/BX}) \times B$$

where

$S/B \times 2^{S/BX}$ = scans per block (general header Bytes 24 and 25)

B = bytes per scan (general header Bytes 20, 21, and 22).

Note: S/B and S/BX are written in the header as binary numbers; whereas B is represented as 6 packed BCD digits.

Assume, as in example E2, that there are 378 bytes per scan. Further assume that S/BX is 2 and S/B is 173. Then $(S/B \times 2^{S/BX}) \times B = 173 \times 2^2 \times 378$, and bytes per block = 261,576.

E10 Determination of scans per block subject to block size limits

S/B and S/BX will normally be chosen to result in the largest block (fewest gaps) possible without exceeding a predetermined limit. This is accomplished by selecting S/BX to be just large enough that $2^{S/BX}$ is within a factor of 256 of the maximum number of scans allowed. Maximum number of scans = integer portion of block limit per bytes per scan.

$$S/BX = \log_2 \frac{\text{Max number}}{256} \text{ rounded up,}$$

$$S/B = \frac{\text{Max number of scans}}{2^{S/BX}} \text{ rounded down.}$$

For example, assume that there are 378 bytes per scan and the block size may not exceed 250,000 bytes.

$$\text{Max scans} = \text{Int. of } \frac{250{,}000}{378},$$

$$= 661;$$

$$S/BX = \text{Log}_2 \frac{661}{256} \text{ rounded up,}$$

$$= 2;$$

$$S/B = \frac{661}{2^2} \text{ rounded down,}$$

$$= 165;$$

$$\text{Block length} = 165 \times 2^2 \times 378,$$

$$= 249{,}480 \text{ bytes.}$$